Veronica's Window

By

LadyHawk94

ISBN: 0-75963-965-5

This book is printed on acid free paper.

1stBooks- rev. 06/18/01

A novel based on romance and the Internet

I wanted to write a book for those who have not yet experienced all the wonders of the Internet. Cyberville, also known as fantasyland, is where you can be anything your mind can imagine. You can discover an ultimate amount of knowledge from the net. No matter what you're looking for, the Internet can be your friend or your enemy. I have often thought that the Internet should be considered the 9th Wonder of the World. Hundreds of thousands of people are online daily. Some use it for helping maintain order in their daily lives, such as for banking or business. Many, like myself, use it as their window to the world. Some, like myself, are seeking other individuals who are home-bound for one reason or another—who are lonely and looking for human contact for either conversation or romance. Yet, I would have never believed that I could become so involved in a soap opera-like lifestyle. My life became a fantasy of romance, mystery, love, heartache, and physical suffering, mixed with a few magical moments along the way. The only problem was that I became so involved in the Internet that the real world disappeared around me. The reality that I was experiencing was only in my own mind—or was it?

There were many people who came into my life that helped me write this book and there were a few who I can always call friend. I will always carry a special place in my heart for them. I love you guys—Gwen, Kate, Linda, and most of all, my imaginary lover, Larry. Needless to say I will never tell a soul which parts of this story is true or my imagination but I hope you enjoy reading it as much as I did writing it.

My dedication page

Where ever we go and for what ever we do in life I will always love you totally with all my heart and soul. Without you I could have never written this book Thank you so much for coming into my life crazy" man.

LadyHawk94

Chapter 1

The E-mail

As I lay in my bed, in a peaceful sleep, I stirred to the sound of the curtains playing in the breeze. Suddenly, I become aware of the softness of my bed, the scents of my sheets. With the window open, I can hear the sounds in the forest as I begin to stretch. Moving my hands over my body from my hips to my breasts, I'm gently pulling the life back into my soul. It seems that all my senses are alive. I feel like a cougar all sleek and smooth, ready for the kill of the night. Seeing your shadow next to me, I shift closer, feeling the warmth of our bodies touching. Pressing myself closer into your manly frame, I tenderly kiss the back of your neck, moving my face against your skin, inhaling your scent. You move slightly, still dreaming, as my hand caresses your side tenderly. The presence of your body is like an electrical current drawing me closer to your firm length. My mouth works skillfully, kissing, tasting your skin; my tongue dances lightly over your responding flesh. My hands glide down your body, sure of their destination. I'm acutely sensitive to your signals and your shifting tells me that you are waking now.

The fire has begun deep inside of our beings. Rubbing your chest slowly, I smile as your nipples become alert. It is time to get your total attention. Roaming, touching your body, I move even closer so that my breasts are pressing against you. Your hands encircle mine, holding them firmly, and I listen as your breathing becomes faster. I lean in to kiss your back, moving my mouth slowly against the warmth of your hot flesh, allowing my tongue to take the lead, making playful rings as I go. Our hearts beat rapidly as you enjoy my wet lips on your skin. In a flash, I remove my cotton gown as you roll over onto your back. I see you are enjoying this as your eyes dance upon my naked body. I cannot get enough of your taste. My mouth dances down your stomach, making certain that you are aware of my hard nipples pressing against you as I go. A soft moan escapes from you and I let my fingers blaze more torrid with each movement.

With one slight shift, you overpower my willing affections, turning towards me so that our mouths lock together. Your tongue reaches for mine; my mouth responds with the contact. You touch my breasts, kissing them, sucking them, sending me into delightful waves of rapture. My hands seek out your manliness, engulfing it tenderly, lovingly, caressing you and then you move again. You are sure of yourself, knowing me as I know you. Inside me, you feel as natural as the rain falling and our bodies unite. We know pleasure is there and we are ravenous for each other. With every move and each stroke, our awareness deepens. I shift my body one final time to confirm the joining I can feel. We are as one now. My hands hold and guide from your back—no time for kisses, no time for words. Your moans and fast breathing matches my own. Feeling the wildness of waiting, I want to scream but hold back as the moment of pure ecstasy begins. I grasp your body intensely, holding you, not letting you fail. I know heaven is close as my own skin tingles in waves. My fingernails dig deep into your back and you respond by becoming the man I love—my hero. The world could end at this moment and we would never notice or care. The rewards now start to lighten and still shaking a bit, our breathing begins to return to normal. You kiss and hold me as we drift back to sleep in each other's arms whispering our, I love yous.

Taking a deep breath and smiling, I push the enter key, sending the letter through my e-mail server. I felt certain that this would stir my Larry to his very soul. Satisfied with my work, I closed my computer down to smoke a cigarette.

I honestly felt like I'd had sex without the presence of another human being. Could this fantasy come true? Even though we had actually never met face to face, we had been talking for months. There was an instant friendship at first and lots of time spent in private messaging. Of course, there were all the phone calls where we talked for hours it seemed.

In our home chat room in front of the world, I was his queen. I fell so madly in love with him. This e-mail I hoped would accomplish two things. First, I wanted him to feel my love; I wanted him to feel like my king. Second, was to show him the ultimate passion I carried for him and him alone. I was so in love with this man that my mind was irrational. I was like a teenager all giggling and unsure of herself. Since we had never met in person, I wondered if it was possible to love someone that much.

It was time to get back to the real world around me. The dog needed feeding and I needed my bath. From the look of my apartment, it was in bad need of some attention, too. As I did the dishes piled into the sink, my mind drifted back to the night that my Larry announced to the whole chat that he was in love with me. Thinking about it made me so happy and proud. At that moment, I knew I would always love him and was on cloud nine as I worked. Larry was now my whole life and I thought I was his. I wanted to be his wife and love him forever, to take care of him and to share whatever came to us. I felt certain there was nothing that could stop us.

Chapter 2

Way before meeting Larry, I'd had several encounters with men I had met from the Internet. I can remember feeling so lonely at times that I just needed to hear another human voice besides my own. MSN, Yahoo—all have chat rooms where people can chat with others online. I met this particular man in MSN chat and found out that he lived in the next county from me. We talked on the Internet a few days and then I gave him my phone number. When he called me, I was shocked to hear the sexy voice on the other end of the line. We ended up talking for hours on the phone. After that, I felt comfortable with the information he had given me about himself. Still, I was so naïve about people and the Internet that I believed too much at times.

Turning every shade of red imaginable while we talk on the phone, I agreed to meet with him. We made a date for the following day and were to meet at my house for coffee. I was so excited about finally meeting someone from the Internet that my mind closed down to all the possibilities of harm that could befall me from this action. Instead, I cleaned and bathed as though I was expecting the President to drop in. I didn't look bad, it was just that I was very picky about my appearance for this meeting. Wanting to look my best, I started pampering myself.

At 5 p.m., the next day, there was a knock at my door. When I opened it, I almost forgot to breathe. There stood a very handsome, tall, sensuous man. The most exquisite hunk of male I had seen in ages. My poor heart was thumping like crazy. I was also blushing for some reason—maybe it was my blood pressure. Finally, I stopped grinning long enough to invite him inside the door. I felt so silly blushing like that. We sat on the couch talking about the Internet and ourselves. We talked about our children, showing our family pictures, and then talked

about the weather and our dogs. All of a sudden, his arm
went around my shoulders, pulling me by my neck towards
him roughly. Next thing, he started kissing me all over my
face. Thinking fast for a moment, I decided that kissing
wasn't so bad, in fact, it felt great. He kissed every spot of
my face feverishly and starting working his way down my
neck. Placing one hand on the back of my head, he held
my hair very firmly—almost too tight.

It seemed like he was really expecting too much for the
first meeting and his kisses started to burn my skin a bit.
Finally, I pulled back as he moved his hand to my breast.
Sitting there, hotter than a pepper patch in July, I still had
some morals. Straightening up and moving back away
from each other, we began to talk more about ourselves. I
was fascinated with his work as a private detective and he
spoke of traveling a great deal. He explained that this was
the reason it would be hard for him to have a relationship
with a woman. He was always on the road and gone for
long periods of time. This suited me; I certainly did not
want a relationship with anyone either. Not anything
permanent anyway. I was coming out of a bad marriage
that had been filled with too many lonely nights already.
Honestly, I think sometimes we can communicate what we
want from the other person without saying a word. We
must have talked for about another two hours before he
said he had to leave, but he asked me face to face if I
would meet him the next day. I accepted without even a
second thought.

My poor body was aching with desire and need. God, I
wanted sex; hot, steamy, hard and fast sex. This tall man
would be my victim. I grinned as I floated through my
apartment still in dreamland. After he left, I went into the
bathroom to get a cool cloth and almost fainted at the sight
of my neck. My poor neck was red and angry looking from
"love bites." Oh dear God, this was totally unacceptable.
Still, I kept thinking that if he made love like he kissed, then

I would be a happy lady tomorrow evening. I saw no dishonor or shame in being human. The night was still young though and I put my gown on and headed back to my computer to see what else I could discover.

Touching a few keys, I was back online traveling through the many links of the Internet. I journeyed into the wilds of several chat rooms looking at all the unusual and outrageous usernames I found. Just looking at these names could make my time worth the effort spent. So, for many hours I just sat in silence in front of my computer giggling at the names I came in contact with. The sad part was that I had forgotten that we humans have the greatest imaginations. The Internet created a positive effect on me because it stimulated my own mind, my own imagination. There was no time I recalled as a child where I had used my imagination for anything, not even in play. Here and now I was discovering what I had never had and I had the World Wide Web to thank for it.

My world had always been harsh, a very brutal reality that I needed to escape from at times. Problem was, I was spending all my time escaping and didn't see the world around me disappearing. My need to travel daily into this fantasy world was getting stronger and lasting for hours at a time. The addiction to being online grew with every waking moment.

As I surfed along reading the articles in my news box, I would leave the chat screen open to watch the happenings in the chat rooms. Out of the blue, I received a PM (Private Message) from a stranger named "JimCee." My first thought was that this Jim was probably a certified nut. Still, what the heck, I didn't have anything to do but kill time. Knowing I was safe in my home, I answered with a simple hello and hit the send button. In a flash, a reply was sent and the next PM made me sit straight up in my chair. If the CIA had sent me a message, it couldn't have gotten my attention any more than this. I read, "Hi, I'm a

6

fifty-three-year-old man who is handicapped and my wife does not believe in sex any longer because of my disability. I'm looking for a real lady to satisfy here, including myself. Are you interested?"

Good grief, I thought, the poor man wanted us to have cybersex. My own knowledge of cybersex was very limited and I didn't understand how someone could type and play at the same time. Besides, I was a real virgin to this game. Laughing to myself, I thought my hands were always too busy when I masturbated. I answered him by saying, "How do you think we can have sex here?" I was curious about the answer I would get. I hit the send button and waited.

A long silence followed before I received a reply. This reply was not from JimCee but from his wife, or that is who she claimed to be. It read, "My husband is a sick old man and I will not have "cyber" sluts after him," she typed to me. Oh my Lord, no wonder the poor man was searching the Net for love and sex with a witch like this for a wife. I was feeling much sorrow for this man but not enough to provide him with a sexual fantasy. I began typing my reply.

"Dear Lady," I typed, "the shame lay with you. How dare you judge this man to the point where you have denied him the joy of human contact and the joys of sexual pleasure. You should open your heart and if he still loves you, then you should thank God every day that you have him. Do not blame him or others for the simple human needs we all have," I wrote. Feeling confident with my message, I hit the send button deciding it was time to clear out of "cyberville" for the night. I had a big day coming up and hopefully, a great evening of romance to look forward to. This strange encounter had taken the fun away from the day, I thought, as I shut down to watch television for awhile. Sometime a little later, I drifted off to sleep on the couch, lost in my own dreams.

The next morning, the routine was the same. I turned on my computer, then started my coffeepot. I could smell the coffee perking as I waited for my computer to warm up to check my e-mail. Waiting in the inbox was mail from John saying that he would call when he arrived into town and let me know which motel to meet him at. My neck was still burning as I replied to his letter stating that I would meet him any place or any time—but no more love bites!

All day I pampered myself. I took a nice bubble bath and did my hair and nails, making sure everything was just right. No matter what happened, I did enjoy making such a fuss about my appearance and I had no reservations about the evening ahead.

The morning passed very quickly and soon my telephone started ringing. It was John. He told me where to meet him and to hurry over and then the phone went dead. Thinking of how strange his call was, my stomach began to churn a little bit. What if I entered the motel room and he decided to beat or even kill me? I recalled how he'd sat in my living room without hurting me, so I had no reason to think he would now. I shrugged the feelings off as I grabbed my bag and headed out the door.

Arriving at the motel just about a mile from my apartment, I didn't have any trouble finding the room number he had given me. After tapping on the door lightly, John opened the door so fast it startled me, but as soon as I saw his smiling face, I began to feel better. With one swift move, he grabbed my hand and pulled me into the room.

I was amazed at the effort he had put into the evening. Incense was burning and there was a bottle of wine with two long stemmed roses on the table. Two very nice wineglasses rested on the bedside table. Suddenly, John pulled me to him and started kissing me passionately, stopping only to pull out a chair for me to sit on while he opened the wine. I had to admit, it was a very nice, romantic room he had put together. The television was blank, but music was playing softly; the lights were dimmed and the room smelled of berries—my favorite scent. I was delighted with all the thought he had put into this moment.

Moving around the room, I sat on the end of the bed and we talked. All the time he was rubbing my arm and my back, keeping my glass filled with wine. Finally, he glanced over to the bag I had brought in and asked what was in it. Opening it up and smiling, I eased out a long, black gown, gently fanning the satiny folds while watching his eyes gleam as he smiled. With that look, I went into the bathroom to ready myself for what I thought would be a grand night. Taking about five minutes, I was careful to check my appearance in the mirror before I walked through the door to John.

His eyes showed me how pleased he was as he reached down, pulled back the spread on the bed, then turned the lights down even more. Smiling to myself, I couldn't help but think that this was much nicer than any honeymoon I had ever heard about. John kissed me several times, then looked at me and said, "Honey, I'm having a problem getting it up tonight." Holy cow! I thought. All this glitter, the wine, and he's going to be a dud in bed. I hid my smile thinking, this is just my luck. So I did what any normal woman would do, I began to kiss him all over his body, using my knowledge of eroticism.

Touching him gently with sure hands, I worked for over an hour, but nothing was happening. Finally, I asked him, "Did you maybe have too much wine?" He responded by saying, "Oh no, this happens all the time, I just need more attention."

More attention? I thought as my mind exploded in laughter. Thinking perhaps it was me, I asked if he preferred something I wasn't doing. He said that I was doing everything great. He informed me that it was just his problem not having sex often enough because of all his work duties, but he wanted us to play longer. Looking at that gorgeous, useless body, I knew there was a possibility that I could grow very old here.

I worked for almost two more hours and his dick was still as soft as ice cream. Still he asked for more foreplay and I responded by sticking my foot right in my mouth. Wide-eyed and tired, I looked at John and made this statement: "I'm not God, I can't raise the dead." In that instant, I knew I had said the wrong thing. Funny thing, I didn't care as I sat up in bed with my mouth drawn tight to keep from laughing out loud. John played around for another hour and then began to mumble to himself. I remained very still and patient but underneath was as disgusted with him as he was with himself. It was no surprise when he stated that he was going to have to make it a short evening due to an early appointment the next day. In record time, he had his clothes on, kissed my forehead, and out the door he went.

Falling back onto the crumbled sheets, I rolled as I laughed. There was no way I was feeling sorry for him because he knew he had the problem before he made this "den of love." Pouring myself another glass of wine and lighting a cigarette, I decided to watch some television. A little later, feeling all warm and snug, I snuggled up to the extra pillow and drifted off into a very peaceful sleep.

Chapter 3

I had surfed and played online most of the night, constantly checking my e-mail waiting for a letter from Larry but there had been nothing in my mail box so I decided to call it a night. A fitful night of tossing and turning in my bed caused me to stumble out the next morning like a zombie. Turning the coffee on, I watched the computer booting up—slow as always. I knew I was being in too big of a hurry and a bit paranoid, but I needed to check that mailbox again. Waiting for me was an answer to the sensual letter I had written Larry. With all the romance he could muster, he explained how much he loved me and how much he'd enjoyed the letter. He said that I was the most passionate woman he had ever met. I knew what passion was, I just didn't have much practice in real life anymore. Smiling, I knew I would travel to the ends of the earth for this man. Yet, something seemed wrong, I could feel it in the words he typed. All of a sudden, I felt he didn't love me as he'd proclaimed. My heart pounded as I could feel him pulling away and I didn't understand why.

A shadow hung over the rest of the day and I was jumpy, almost in a panic. Something was missing in his letter, or perhaps it was the wording, I couldn't be sure. The thought came to me how strange it was that a man would love, fight, and die for his country if needed but for a woman's love, would run like a scared rabbit or lie about it. Maybe my imagination was working overtime. Brushing aside my apprehensions and calling myself foolish, I planned the day ahead. It was basically like every other day; the only light of the day was when Larry called me on the phone. For this reason, I'd planned everything so that I could be home to catch his call. I could call him and often did, but I knew he was busy working on some cars so I tried to leave him alone. That day went slow, hour after hour I waited and no call came, no other e-mail either. I was almost a basket-case trying to keep silent and not worry. I said nothing to my online friends about my fears. My thoughts tried to stay rational as I kept thinking anything could have happened. The power could have gone out or his ISP server could be down to prevent him from getting in touch with me.

As the day turned to evening, my hands started to shake and I decided that I would take a bubble bath and light some candles to help me lighten my mood. Before I could get in the tub, my supper decided to make a return trip and my head roared from being sick. Picking up a bottle of strong pain relievers, I shook two out into my hand, swallowed them without water, and climbed into the tub. Whatever was happening, I knew I would need help in dealing with it and at least the drugs would ease my head and stomachache.

I had been involved in a traffic accident earlier that year that had left me with a ruptured disc in my lower back. To make matters worse, this was the third time it had ruptured in the past two-and-a-half years. My lower back was a mess and my legs no longer worked well. Sometimes I felt like I wasn't worth much anymore and didn't have the nerve to ask anyone to be burdened with a cripple. My doctor didn't want to operate again until the pain became unbearable, but little did he know that I had hit that point many times during the day. He wanted it to be a full-blown rupture again before he moved with it. With this rupture, however, he prescribed stronger pain medications upon my demand. I devoured large amounts of those pills and sometimes, I couldn't tell the difference between my heart breaking and the actual pain in my back and legs. My moods changed with each passing hour of the day.

The funny thing was that I wasn't aware of what was happening to me either. If you would have ask me if I was addicted, I would have laughed and said no. In fact, no one around me knew what was happening; not my kids, not my friends, no one. In the real world, my friends never saw the problem either, then again, I didn't have but one or two friends that saw me on a weekly basis.

As I look back, my computer had become my life and I used it as one uses a friend. Continuously, I sat in front of the machine, day after day, hour after hour. I joined online clubs, from gardening to health, clubs that paid for surfing through online commercial ads. I tried many things to take my attention away from my real thoughts and pain. I was losing the battle and unaware of the war.

Over the next couple of weeks, I would see Larry's name light up in our chat room. Each time I would try to talk with him, he would only respond with very short, direct replies. The people who knew we were an "item" would ask me what was going on and I would break down. I had no answer to give them because I didn't know either. He knew I was physically hurt, but never offered any kind words or a shoulder to lean on. There were no words of comfort or love, and I was falling apart. My emotions were getting out of hand and there was nothing I could do. I asked him many times what was wrong and always the same answer—nothing. Yet, the phone calls stopped and he never said anything to me unless I sent him a message first. I was constantly going over everything in my mind trying to figure out exactly what I had done wrong. Each time I came up with the same answer. I had done nothing wrong but I carried the guilt as though I had.

Some of my friends suggested that Larry was a player, but I refused to believe that and started sending e-mail after e-mail without any response. I was sick with despair and hurt. Time seemed to crawl along for me. My mind was spinning uncontrollably and I decided to drown my sorrows in the wonders of someone else's mind. It was so obvious that Larry did not want me. I took more pills, wanting the drugged state to disguise the true pain I felt inside.

Sometime later in the week, I went looking into the personal ads on Yahoo. I was just looking at first, but then I came across another man who lived in my own hometown. The ad sounded sincere and after some thought, I answered with a short hello, adding my e-mail address to open the line for communication. Within a few hours, I received a reply from Rex. He was a business owner, fifty years old, divorced, and lonely. Since we lived in the same town, it was easy to set up a meeting at the local Waffle House for coffee. We set up and agreed on a time to meet. Never in a million years would I have guessed what was about to happen.

When I saw him, there was no doubt as to why he didn't date much. I was no raving beauty by any means but this man had three stomachs too many. Looking very unkempt in his gray suit, I decided I didn't care. I just wanted to talk anyway. My mind wasn't even thinking about having any kind of torrid sex with this man, just human conversation.

After we talked for a couple of hours over coffee, I decided that it was safe to invite him to my apartment for coffee again the next night. Nothing about this man turned me on sexually, but my need for human contact encouraged me to offer him an invitation. Friendly talk over coffee or tea was refreshing and needed now.

The following night, I did nothing special to prepare my apartment as I would normally have done. He showed up on time holding freshly ground gourmet coffee to find me in a state of disarray. I was wearing my regular clothes, which were nice enough for a coffee date.

Everything went fine at first. I made the coffee and we talked more while I made myself busy setting spoons and cups out. We talked and watched a movie on television; I even popped some popcorn for munchies. Just as the movie ended and I was about ready for him to say good night, Mr. He-Man grabs me, twists me around, and pulls me down on my couch. I was under him faster than lighting could strike. There was no way I could move any part of my body. Somewhere in the back of my mind I couldn't help but to think that he had too much knowledge about this act to not have done it before. He was smooth and knew what he was going for; a real master at pinning someone down.

My back was in pain as I cried and I couldn't breath with his weight suffocating me. The room started spinning and I thought I was going to pass out when he shifted and at last, I could inhale again. My voice was telling him to move off my chest. Panic swept over me because I couldn't breath while he held me even tighter. At that instant, I knew I was about to be raped—the man had already shown me his strength.

I am going to be raped right in my own home, I thought, over and over again. I told myself to keep calm and to keep my head straight, not allowing myself to lose control now. Fighting could make everything go bad and cause him to hurt me. Talking to him only enraged his passion. His powerful grip prevented me from moving in any direction to free myself. Mentally, I was thinking that I deserved this for being so stupid.

With another forceful twist, he held my arms behind me, unfastening my clothes as he mumbled about how beautiful I was and how my breasts were gorgeous. My body, he could take, my mind and soul I would let escape to a far away place, not thinking anymore, not feeling. It took only a couple of minutes for him to strip me and relieve his sexual excitement on my body. Repeatedly, he tried very hard to enter me, but could not. Focusing my mind, I had contracted my muscles so tight that under

15

normal circumstances, a man might have thought I was a virgin.

I thought I would be sick to my stomach at the disgust of it all, while I lay silently waiting for this assault to end. When he was finished, he rolled off, dressing hastily, then rushed through the door. My mind raced as I headed for the shower, throwing up my dinner and all the coffee we had drank. I would not call the police because of the shame I felt, but my mind thought of revenge. For the next few days, I lay in pain from my back being twisted and planned my course of action. One good thing about living in a small town, it is difficult for people to hide. The saying, hell hath no fury like a woman scorned, was something I was learning intensely.

Chapter 4

For the next three days, I plotted my revenge. On the fourth day, I was waiting for Rex when he came out of his small office complex. I saw him walking toward his car in the parking lot. As he drew near, I opened my car door, calling his name sweetly. "Rex, honey," I called out and he looked over at me. When you go bear hunting, you carry bait, and my bait was a short skirt, heels, and a very low cut blouse. At first, he had a nervous look but I flashed him a beautiful smile as I waved and that got his attention. "Hi, honey," I said with all the phony southern charm I could deliver and he approached me, smiling. This man was not only a "rapist" but he was so arrogant, too.

Flashing him a sexy smile, I told him how much I'd enjoyed his rough actions at our last meeting and asked if he would come back over Friday evening and do it again. I grinned like an innocent schoolgirl while batting my eyelashes his way. The idiot fell for the bait. We stood for a few minutes talking beside my car, his eyes never leaving my breasts as we spoke. I knew when I'd bought that blouse that it was a sexy number and his face told me I had been right. The blouse would be burnt after this was over. Laughing and swaying at him, I got back into my car. All the time Rex was drooling like a mad dog over me. "See you tomorrow night, honey," I said to him, then drove towards the hardware store. There would be a few special items needed for this party.

The next night, I had everything set up when Rex knocked on my door. He arrived on time carrying a large bouquet of flowers for me. Inviting him in, I asked that he make himself comfortable as I went to the kitchen to get a vase for the flowers and to get our coffee. His coffee would have something extra added—three ten-milligram Valium tablets. I smiled as I stirred the drugs into his cup.

Almost laughing to myself, I talked the entire time as I set the drinks on the coffee table. Taking a few sips of my coffee and watching him drink from his cup, I told him I'd had a very busy day and asked if he would be okay if I took a shower. Smiling at him, I told him to enjoy his coffee, giving him a sexy, come-on look.

I slithered like a snake as I disappeared into the bathroom. There I had a book waiting and I turned the shower on so he could hear the water running. Sitting on the toilet, I smoked a cigarette and smiled. I waited a full twenty minutes before coming back into the living room. There he lay, in all his grand glory, on his back snoring like a pig. "Rex," I said, shaking him, but he was out like a light. Laughing out loud to myself, I changed back into my slacks and blouse. The fun was just starting. My plan was working out well. Picking up his hand, I checked his pulse and it was still strong so the Valium hadn't killed him, I thought, as I dropped his hand.

First, I removed his clothes and threw them in a bag. All the time I was removing his clothes I was talking to him and not in a nice way. I told him that he was only getting what he deserved. Next, I bound his hands behind him with duct tape, then taped his feet. Taking a hammer, I broke up twelve light bulbs into thousands of little pieces inside a paper bag. I hummed nursery tunes and smiled the whole time I worked. Giggling, I pounded the glass a few more times just for luck. I kept a close eye on Rex to make sure that he was still out cold; he continued his loud obnoxious snoring. Finally, I maneuvered him on the couch to a better position. I carefully poured all four tubes of super glue I had bought at the hardware store over his little dick and balls. I made sure no skin was showing—I had even put a sheet under him so I wouldn't get anything on my couch. He did not move or twitch as the cold super glue touched his bare skin. I shook the broken glass over the treated areas with so much precision that it looked like a fine piece of modern art. It looked like a glass diaper when I finished with him and I was pleased with myself.

Almost done now, I thought, as I sat with my blow dryer, going over the area to make certain it hardened well. God I am evil, I thought, as I whistled, aiming the hair dryer to just the right spots. I placed a call to a couple of my old friends who were standing by all excited to help me with my plan. When the two big guys arrived, they laughed hysterically at what they saw, both saying that they were happy to help but muttering comments under their breath about never wanting to piss me off. Hell, I had even held the cheeks of Rex's butt open, pouring in the glass and super glue, filling in his crack. I'd made sure that the super glue surrounded all the hairs to make it stick better. I couldn't help but laugh at my evil deed. My two big friends began dragging Rex out of my front door, one on each arm, when I stopped them.

"Wait," I said, grabbing a black magic marker. I wrote three words on Rex's back in large, bold letters. **Rape Victim's Revenge**! With that done, they took him to a truck outside, threw him in the back, covered him so he would not be seen, then drove to the front door of his business. I was amazed that the Valium had worked so well.

Following close behind the truck, I drove his car back to the parking lot, leaving the keys in the ignition switch. After all, I wasn't a car thief. I watched as my friends placed him against the door of his business and laughed. Oh yes, Rex unknowingly paid my friends for delivering him back safely. Rex had two hundred dollars in his billfold that my friends relieved him of for their kind assistance. I was tempted to feel sorry for Rex lying unconscious against the door with his naked body wearing all that crushed glass. That feeling disappeared instantly as a pain ran down my back and I laughed as I got into the truck with my friends heading home. I set my alarm clock that night before bed. I wanted to be there thirty minutes

19

before Rex's business opened. There was nothing in the world that would cause me to miss the show that would be going on in the morning in front of Rex's business. I slept like a newborn baby that night and felt darn good about my accomplishment.

Morning arrived and found me sipping coffee inside my car watching as employees began to arrive for work only to find Rex reclining against the door of the business in his new suit. I watched closely as Rex woke to find people standing over him removing the tape but not daring to touch the glass diaper. Laughing hysterically, I nearly busted a gut as the ambulance arrived. As the paramedics were lifting him up on to the stretcher, the small crowd of bystanders that had gathered saw the message I had written on his back. The revenge was complete for me when they read, "Rape Victim's Revenge." As the ambulance pulled away from the curb, I sat there smiling, happy with the work I had done and ready to head back home.

Keeping a sharp eye on the local newspaper the next few days, I watched for any news about Rex. Finally, on the fourth day, it appeared. A small article well-hidden in the back pages told of this poor man's dilemma, how he had been a victim of a robbery plot and was undergoing surgery for the cuts on his body. Rex for some reason could not recall what had happened to him. In his statement for the local newspaper, he said he saw some older teenagers in the parking lot of the business and that was all he could remember. Laughing, I laid the newspaper down and took my coffee to my computer. Life was so good at times.

Feeling renewed and at the top of the world, I decided to write Larry another e-mail. At this point, I did not see that it could hurt matters. I had nothing to lose and besides, it made me feel better. He was avoiding me for reasons I still did not understand and nothing I did or said made any difference to him. With everything that had happened, I still missed him and only wanted to understand what went wrong even after all the time that had passed. I wanted to stir him into some kind of response and, at this point, any response would certainly be better than what I had gotten from him so far. Typing the e-mail very carefully, I thought of him. In my head, I could see and feel his body next to mine. I tried not to think of myself as foolish but as a determined fighter.

My dearest darling, I made a wish last night as I lay in my bed. The wish was to have you with me, to feel your naked body against mine as we make love. I clutched the pillow beside me and my thoughts were of you. My body quivered and my mind ran into visions of sexual pleasure. With our naked bodies pressing together, our lips grow so hot that we tremble when we touch each other. I glide my body over yours, moving smoothly and sure of my actions. While our hands are stretched out to the sides, I kiss your face, eyes, and neck gently, caressing your chest with my lips. Skillfully, I work my way down your body, tenderly kissing and tasting your skin. My tongue flickers across your nipples, feeling them growing harder. While you moan loudly, shifting your body closer, I move downward. My hands touch every spot as my hot mouth follows and my educated tongue runs along its own path. I linger as I kiss your stomach, your hands rubbing my shoulders. I move to your thighs, kissing the inside of them, licking and sucking your skin gently—not to cause you pain but only to give you pleasure. I massage your cock with one hand and manipulate your balls with the other as if I were preparing a fine meal. There is no doubt from your hardness and your breathing that you are nearly ready for me. Running my tongue up the shaft of your penis repeatedly, I feel you grow harder and fuller. Stroking and sucking harder, I hear your breathing becoming spasmodic as I slip my hot, steamy mouth onto you fully. My tongue works around the head, coming back up and going down on you a little bit further each time. Gently, yet firmly, closing my lips down around you, you push deeper into my mouth. My tongue teases you with little butterfly circles around the rim of the large head. Your body responds, moving in time with mine as you call out my name with a deep moan, setting me on fire. Your hands hold the back of my head firmly as you whisper, telling me you are almost ready. "Oh my God," you scream as you push me aside, leaving me gasping but very excited. Exchanging places, you frantically kiss me all over my face while rubbing your body against mine. I feel your desire as you take each breast individually into your mouth, leaving no part unattended. I cannot help moving and moaning as you give the pleasures I seek. You slowly work your way down my body making me crazy with desire. The moment intensifies when you push my legs open wider for your complete access. Your mouth kisses my mound, parting my vaginal lips while you bury your face within me, your tongue playing. God, I love you so much. You make me feel entirely wanton as I scream, "I want you in me now." You slide your cock inside me, making me gasp even louder. Every move, every stroke brings ecstasy closer as we pleasure each other. The moment turns from passion to pure lust as we throw ourselves at each other. Your hands hold you above me. My hands hold your hips with a tight grip, keeping the rhythm in line. You call out, "Oh darling" in a loud whisper. This is my signal; it is time to release everything we have worked for. The pounding of our bodies against each other knows no limits. We hear only each other's heavy breathing and panting as we climax at the same moment. Time stands still for us and what seems to be forever is, in reality, only seconds. I could not care about anything in the world. You and I are the only things that matter now and total release and pleasure belong to us. Your body goes limp over me and still locked together, we turn to our sides, facing each other and drift off into a peaceful sleep wrapped in each other's arms. No cares, no worries, just us and that is all we need.

Now, that should do it, I thought to myself smugly. Adding a personal note to get in touch, I sent the e-mail into cyberspace.

The rest of the night went by slowly with me constantly checking my e-mail. There were no messages from Larry. Nothing venture nothing gained, I thought. Needing to hear his voice, I tried calling him on the telephone but only got his answering machine. I left message after message asking him to please call me, saying that we should try to be friends if we could not be anything else. Panic swept through me like fire again. I had lost so much in this world and I was not sure I could handle another loss. Still, there were no answers and no acknowledgements online at all.

Why is it that when you love someone, your fears overcome your common sense? Had I scared him with the first sexy e-mail? He did not call nor send me anything with any type of closure to what I still thought of as "our relationship." I was left in limbo as to what was going on and I hated that. Then a letter finally came from him, but all it said was that enough was enough and for me to please stop the e-mails. Damn it, I wanted to know why, and I needed to know for my own peace of mind. How could he have proclaimed to love me so and then not want to work it out? I was so confused and hurting. I spent the rest of the night crying into my pillow begging God not to let this happen to me, but God must have been on vacation that night.

Chapter 5

Saying I am a true Leo lady often explains some of my high temperament and moods. Of course, it could be my Cherokee Indian blood, I have always said it gave me passion in life and for life. Only Larry had ever called it, "attitude" but what did he know? I was the type that never hurt anything, not once in my life had I ever hurt an animal or another human being. Although I was very aggressive in parts of my life, I was very passive in others. My heart had always been on my sleeve and easily crushed but this time, it was torn up into pieces. I had lived a very solitary lifestyle for many years thinking I had grown too passive in my older years. Yet now I could not help fight the rage running through me. I blamed myself for trusting and trying to love again. Still in a daze, I wrote him what I thought would to be my last e-mail telling him I wanted nothing else to do with him and that he was a jerk. The moment I pushed the send button I regretted it, but I took the defensive and so did he. The relationship was off according to the e-mail he sent me and I knew there was nothing I could do to change it. Is it a relationship when one partner would not speak? I wondered.

Trying to be brave, I sent another letter to him that said that it was obvious that he no longer wanted anything to do with me and that I was releasing myself from any promises made to him. My pride was speaking for me with that e-mail. It was another drastic mistake and a total lie on my part. A lie I would regret for a lifetime.

The next several months found me trying to undo the mistake I had made. My obsession with him grew stronger with each passing day. I could not control myself any longer and I was spinning into emotional desperation. It was not in my nature to give up on something I cared about. Realizing that I loved him more than I had ever loved anyone, I tried to bring him back to me and to bring back the beauty of the feelings we had shared, but there was no bending on his part. It would be months before I heard from him again. I finally realized that it was over for him the night we were both in the same chat room and he asked another female chatter to meet him for coffee. My heart broke. Of course, his asking the other lady to coffee was the ultimate insult for me, the rudest one also. His invitation to her was a direct slap in my face since he could have asked her in a PM and not in the open room. He knew I was there, could see what was happening, and could read what was being typed on the screen. The lady he asked to meet him for coffee never responded as far as I could see, but then, I could not see much for the tears running down my face.

Depressed and angry with myself, I decided to go surfing into the adult chat rooms under another username. I was looking for trouble and I did not care how it came to me. Within minutes, I was learning the fine art of cybersex. I felt so low, death even crossed my mind, but I was broken-hearted, not crazy. This depressed feeling made it easier for me to slip under the rock I had dropped on myself.

One good thing about Internet cybersex is the personal interaction with another human being—although it is only in PM, typed form. The other person not seeing you is another reason why cybersex is so popular. Unless, of course, you are using cam recorders and I know that some do. It's a matter of personal taste in what you desire and how much you really want to interact and see. All I wanted was to bury my feelings and not have to deal with them anymore. I was typing sexy words on the screen and it meant nothing to me. Some of these men, however, could type extremely well and construct sentences that could heighten one's awareness with just a little imagination. I sat in front of my monitor crying like a baby as I played the part of a sex kitten. I typed the words of sexual excitement to a stranger thinking how sick I had become since I'd gotten online.

Suddenly, the screen stopped and then the stranger asked if there was anything that really turned me on. Little did he understand that I was not turned on at all. I typed back onto the screen, "I need to see the words I love you," knowing how strange that may have seemed to whoever was on the other end. My cyberpartner paused for a minute, then typed across the message screen, "I love you, darling." I cried more as I rewarded him with what he wanted to read. Later that day, I would hate myself, but I would learn from my actions.

Weeks went by and I was still feeling very depressed. My back pain had increased and so had the potency of the strong pain pills. I decided to call an old friend, picking up the telephone to dial the familiar number. An answering machine picked up and I left my message. Sweet and simple I said, "Please come over, now is the time for all great men to serve their country." I giggled, knowing he would understand the message completely. Gary would laugh but he would call and I knew it would be soon. Laughing at the silliness, I logged online to surf the Internet while I waited. I was reading an impressive article on natural herbs when a knock came at my door. Opening the door, there stood the world's most lovable guy. He was tall and handsome and he was a great lay. Smiling as if he had just won the state lottery, he handed me a bottle of blueberry wine, then gave me a hug.

"Gosh, that did not take long," I said, grinning at him. He had been across town when he picked up his messages from his cell phone. He'd stopped, bought the wine, and headed straight over. He knew my apartment like his own because we had been friends for many, many years. I knew he loved me, too, and he knew that I loved him as a friend. Ages ago, Gary and I decided that our best bet would be to remain friends only, but to have sexual flings on the side. Who better to have sex with than a good friend? Gary made himself at home opening the wine as I turned off the computer to have a quick shower. Coming back into the living room a few minutes later, I had a towel on my head and was wearing a loosely tied robe. Gary looked at me with admiration. "You are the woman I dream of," he said.

"Oh you," I said as I laughed at him. It was something he'd said all the time when we were dating. I sat on the couch wanting to cut all the small talk short because we both knew why I had called him. After about five minutes of small talk and a half glass of wine, Gary's arms went around me. He whispered as he kissed my neck, "Veronica, let me make an honest woman out of you." He would have married me in a heartbeat but I was afraid that if I married him, it would spoil a great friendship. I smiled at him, giving him my "come get me big boy" look and the games started. I had no guilty feelings with Gary; all I felt was the desires of a woman's needs. In all my years of having sex, Gary had always been one of my better memories. Nothing much needed to be talked over as he opened my robe, fondling my breasts, kissing me all over with the hottest lips in town.

Gently pushing me back onto the couch, he rubbed my legs softly, kissing them tenderly. Gently pushing my legs apart to find what he was seeking, he carefully mastered me like a champion. He kissed my mound gently, working his skilled tongue into the depths of my body. Gary was the only man I knew who got off on giving oral sex, and he was the best I had ever known. Within a few minutes, he had me moaning; he had the best hands for massaging I had ever experienced. If I had not known better, I would swear his tongue grew to eight inches with each thrust in me. He always knew where to go, where to stop, and when to go again. His wonderful tongue darting in and out

27

had me reaching untold depths of climax. Soon I was in the danger zone, and the danger was me screaming too loud and pulling his hair too hard. As I began to cum, he joined me with one last hard, deep thrust of his tongue. Later, he laid his head on my stomach and I rested, leaning further back on my couch, taking a short nap. When I opened my eyes later, a soft blanket was covering my nude body and Gary was gone. This was a normal thing for him to do. A little note on the coffee table said, "You're the best, I love you always." Smiling, I tossed the note into the waste paper basket and decided to get dressed and go shopping. I felt renewed and happy with myself. Happy I had someone like Gary as a friend and sometimes lover.

After shopping and getting my housework done, I turned the computer on. Roaming the many chat rooms of Yahoo was something I did at least three to four times daily. I knew I had become a serious Internet Junkie. My newest adventure was telephone sex and yes, the Internet had opened the door for that experience, too.

It happened one night as I was surfing and "popping" pain pills. I had stopped in a voice chat room and was listening to the talk when I heard the voice of a man that actually sent tingles down my back. Hearing him speak was almost as good as having sex. Leaning back in my chair with my headset volume turned to high, I closed my eyes and became lost in the voice. The ladies were after this man, or could it be that great voice they were after? I giggled as they tried tempting him to talk with them but he was playing coy. They seemed almost desperate to have someone call them. It was sad, yet funny, too. I heard my PM alert ding, and opened my eyes to look and almost fell out of my chair, for there in my PM window was the man with the wonderful sexy voice. He was asking me if I would say hello to him in the open voice chat and I heard him calling my name. Oh, I thought, he wants to judge all the prospects in the room, which seemed smart to me but I was just listening and did not actually intend to do anything in an open voice chat.

Wow, I thought when he said my name in that beautiful voice that was almost hypnotic. Adjusting my microphone to talk, I thought, what the hell, and spoke softly back to him. For some reason, the men gasped and the ladies in the room went dead still. No one was saying a word as this stranger told me I had the sexiest voice he had ever encountered online. I never thought of my voice as appealing. It must have been my soft, southern drawl that attracted him. To my amazement, Mr. Sexy Voice had chosen me as his chat partner. Blushing but also flattered, I could hear him saying all sorts of luscious things to me, in open chat no less. Glancing up to the screen, I saw his PM asking me if he could call me for further conversation. There was no way I was going to call him or give him my number, then he suggested I get a calling card and he would give me his number so that I would feel safe. We would try tomorrow night if I was ready and willing. Since I did not like feeling pressured, we made an online date for the next night at the same time. This would be a new experience, I thought to myself. At this point, I was willing to do anything to kill time and memories.

The following night, I discovered telephone sex and it was great. A nice voice and a vibrator are all the equipment you need. My mind and body were ready when I called Bob. I was very excited, sitting on go and rocking on ready. Lying on my heated waterbed, I made the call to him. He talked briefly about small things like his real name, where he was located, and that he was not married but loved phone sex. Then he started talking of holding and kissing me and I talked of the many things I would love to do to his body. He wanted me to talk of oral sex and I had no trouble saying anything to him for some reason. His voice and my toy were enough to bring me multiple orgasms. When I started to moan in my excitement, he became so worked up that I could hear his breathing getting deeper and faster. This all took place in about ten minutes. With me using a calling card and having his number to call, I was as safe as a baby in mother's arms. When we both could breath again, he said that I was wonderful with my sexy southern voice and that he had cum harder than he had in months. We said our good-byes and I threw the cordless telephone to the other side of the bed ready to sleep for a change. I did promise to call him again, but I never did. A man with a voice like his is the type you want to bring home to keep and that was not in my forecast anytime soon.

Besides, who needs to be out in the real world when you have all the stimulation right here online that you could ever need? My mind had become so stimulated from the Internet that I began to require very little sleep. I told my friends, repeatedly, that I would sleep a long time when I died, but my eyes were wide open now. Sometimes I felt as if I was running amuck and then other times found me still looking for a hero. It makes no sense in explaining my meaning of the word hero, since everyone has an individual opinion of what a hero is. Yet, I dreamed of having my own hero. My own thoughts were ones of total devotion, love, honor, and trust. I was afraid that I would never find one and in fact, I was not looking anymore. My heart seemed dead and no matter what I did, there was nothing to wake it up.

Watching the people in the chat room talk to one another, I thought about all the wonderful, loving humans I had run across in these rooms. Laughing at all the memories of my online adventures, I started getting PMs from my friends about this and that. I enjoyed anything that would consume more of my time.

I had been online long enough to have made several very wonderful friends—two ladies in particular that I had met in a chat room. We starting chatting one day in the 60s rooms of Yahoo and have been buds ever since. We played music, worked on computer problems, and chatted everyday together about ourselves and life in general. I am proud to call them my friends today. The Internet is not always a bad place to play, but one should use common sense while visiting there and chatting with strangers. Not everyone is kind or without evil intentions.

This day online would be no different from the rest. Sitting in chat talking to friends, I received a PM from a man who said he had been watching me and had decided I was going to be his. This type of message is the kind to take seriously. You never know if the person is a stalker or just some everyday guy who is trying to get your attention. It could even be a woman pretending to be a man. I played it cool and replied, "Why do you want me?" After sending that message, I joined him in a voice chat and decided that if this was a stalker, I needed to know as much as I could for any future problems. I went into voice chat and we begin talking almost like old friends. Something very familiar about his voice kept tugging at my mind. I thought he sounded like Larry but I could not be sure. I let him talk as much as possible, not answering any questions about myself. The more he talked, the more he sounded like my Larry, but he said his username was Ed. It had been months since I had actually heard Larry's voice, but one thing you never completely forget is the voice of the man you love.

This man claimed to be a player and wanted to know if I wanted to play. Player is a term for anyone who is online for fun and games without commitment. I decided to be a player with him to see how far he would go with the game in case it really was Larry. The more we spoke, the more I became convinced that it was, but I would take it easy with

31

him just to see where it would lead in the days ahead. I would never make it easy for him or anyone to hurt me again, so I played the game, and I could play it well. Just thinking there was a possibility that this Ed could be my Larry was enough to bring out all the psychotic and paranoid feelings of the past again. All the hurt feelings I had worked so hard to bury now resurfaced. I kept saying to myself that I was okay with talking to him but in reality, I was not. My girlfriends were truly worried about me. I even doubted my own reasoning for wanting to talk with this Ed guy. Still, my friends wanted me to be careful.

I grew more certain with each word that Ed spoke that it was Larry. Maybe I just wanted it to be Larry so badly that the truth simply fogged my mind. Kate and Gwen stood by me and gave me all the encouragement I needed to pursue the matter and try to discover the truth about this Ed character. I felt sure I had a tight enough grip on my emotions that I could deal with anything, but that was far from the reality. Many things Ed said to me were exactly like the things Larry had actually said before. I found myself catching my breath at times, sometimes even trembling from fear. I began taking notes on things he spoke of and noted how his actions were like Larry's, too. Larry had been the only one who'd ever said I had an attitude and one night in private chat, Ed said the very same thing. When I asked him to explain that statement, he said never mind and went on with the conversation. This not knowing was driving me nuts all over again.

With each conversation, my heart was ripped open and I was not sure what to do. If Ed was not Larry, I did not want to talk with him, but on the other hand, if he was, I wanted to stay as close as I could. Deciding to play it cool and not rush, I would bide my time and see where it all would end. I did have faith that one day, everything would work out but how long it would take, I had no ideal. Time was not really on my side anymore and I wanted to love and be loved before I left this world.

In the meantime, I had met an older man from California when I'd first come online years ago. He turned out to be a very nice online chat friend. There had been many nights where we would spend hours in private messaging. We had become very close as far as online friends and I had given him my phone number so he could call anytime he wanted. He was a jokester and always a good time charley but he was never rude or vulgar. He would always come online with a joke to make me laugh. At times, I would discuss my problems with him, asking for his opinion. No matter what I said or told him, he was never shocked nor did he ever judge me. For this reason, I grew closer to him as a friend.

One day while I was on the Internet checking some music sites, I received a PM from him. He asked me if I wanted to come out and visit with him. Oh, a real country girl in California, now that was a thought to boggle my mind! He said that if I wanted to come, he would buy me a round trip ticket, that he had an extra bedroom I could use, "No strings attached." I felt totally safe with him but decided to think it over; besides, my feelings about men had not been very accurate within the last few months. The thought appealed to me because I had never been out West and I had never flown on an airplane. Yes, I was forty-eight and had never been on an airplane. However, I wanted to think it over and maybe get some more details before I made up my mind on my friend.

Answering his PM, I said, "I appreciate the offer and will think it over." The idea of me getting on a real plane just blew my mind. Everything changes in the blink of an eye for me and I was always ready for the next change—or tried to be. I kept praying for good changes. When I was younger, I was a leader, running into any situation head on—damn the torpedoes, full speed ahead—but I was older now and those damn torpedoes were getting the best of me.

My life seemed like a never-ending circle of turmoil. My very best real life friend once told me that she loved to talk with me because my life was constantly moving in different directions; it was like a soap opera on television. When she told me she admired my strength and my openness about life, it made me feel better in some ways but I was tired of the soap opera. My search was for peace and tranquillity now. Years of abusive living and treatment had weakened my heart and forced me to sit in the back seat at times. I always had a sense that I would not live to grow old gracefully. Major disasters all through my life made me feel like it was God's way of telling me to live fast because I would die too young. My life seemed unreal at times with so many things going on and 98% were from my adventures online.

Chapter 6

This particular morning found me playing in my home chat room when the telephone rang. It was an old friend from the Veteran's Hospital informing me that Ronnie Lane had passed away and asking if I wanted to attend the service Sunday morning in Memphis. Of course I would attend. I remembered how Ronnie and I had developed a friendship based on trust and understanding. Several years back when I'd worked as a nurse, I'd done volunteer work at the local Veteran's hospital every other week. Writing down the details, I felt myself starting to shake from the news.

Turning my computer off, I sat down on the couch and cried like a baby. My mind started pulling up old memories of Ronnie and I talking, many times about life and death for that matter. Our walks through the gardens at the hospital seemed to make us both feel better, too. The stories he told me of Hamburger Hill and the things he'd experienced and saw while serving his country made my blood run cold as it always did. I could almost hear him talking to me, laughing in that wonderful voice he had.

Ronnie had been a very special friend. The first time we met, he had just arrived at the hospital for his first chemotherapy, a tall man but very slim. I could see from his innocent look that he could not harm a fly unless forced. I knew that he'd left the military with the highest of honors for saving lives in Nam. We must have talked and laughed for hours. Each day I looked forward to seeing him, bearing in mind of course that I had to distances myself from the men I talked with. There were constant warnings from the staff personnel officials against forming attachments to the men, which was understandable. Ronnie seemed like a long lost friend the very first day we meet. In one conversation, he told me that his wife had left him some time back and there were no kids. He blamed himself for not being settled enough when he returned from the war. Now he faced cancer alone and that had to be the worst feeling in the world. Still, I enjoyed listening to him and we even laughed at times about his skin coloring, which was a dull orange, but I never asked any questions. I was there for him, listening to his dreams and thoughts. Mostly when we talked, he would tell stories of a far away country, which sounded beautiful but had taken so much away from America. He told of many sons and daughters fighting a senseless war and dying right in front of his eyes. When I saw that certain faraway look in his eyes, I would change the subject to something more pleasant. He promised to teach me fly fishing one day. I would laugh at him saying that I might not be smart, but I knew that fish couldn't fly—he would laugh so hard. It was our private joke.

One morning when I arrived at the hospital, I found Ronnie lying quietly in his bed. Walking softly into the room, I saw his eyes closed and turned to leave. "No, please do not leave me," he said. Pulling a chair close to his bed, I sat down as he started talking in a low, weak voice telling me he did not want to die yet. My heart was breaking as I held back my emotions while holding his hand. His firm grip on my hand told me how frightened he was as he spoke in breathless whispers. With each sentence from him, I could feel the tears bubbling up inside me. I rose from the chair and eased down beside his warm body. I could feel the cool sheets as I lay quietly holding him. We both cried silently, never saying a word until finally, I knew he was sleeping and then, silently, I moved away, kissed his forehead, and left his room. Looking back now I remember speaking to God with my request, "Take care of him please." I prayed hard as I closed his door and left. This would be the last time I would see my friend Ronnie alive.

Being emotionally overwrought, I stayed away from my volunteer work for over a month. I could not emotionally handle it any longer. There was too much death and too many broken hearts and I wanted to help everyone but I could not. When I did return, Ronnie had left with no record indicating where he was going. My mind told me if he was well enough to leave then he was okay, but I knew better. Still, Ronnie and I kept in touch over the next few years with Christmas and birthdays cards. I had given him my address when we'd first met. Every time I received a card from him, the address was from a different Veteran's Hospital. I always hurried to return his mail because I knew he would move again. Hearing of his death brought tons of hidden memories back. On Sunday, I would stand at his grave bidding him farewell one last time, but for now, I needed a big drink, hell, maybe two drinks. I needed the Net to busy my mind so I headed out surfing.

Online I went to my bookmarked pages on American Veterans and of course, the music on those pages was always very beautiful. I had always been a sap for my country and loved it. Often there were times in my youth that found me wishing I had been born a man so I, too, could have served my country. That wish was never fulfilled although I tried to join the navy to be a navy nurse. My heart would not pass the physical even then, but I supported the American service men with all I could do here at home, mostly with prayers and with my volunteer work.

I spent the morning remembering him, wishing his outcome could have been a happier one. My rage seemed bottled up and would not surface in me but I played hard with my sexual side, much harder than most of my friends had ever seen me play. LadyHawk94 is my chat room alias and I was always a lady in the room, never allowing anyone to know the truth about me being a player. This time when LadyHawk94 was in chat she would play music and shake her tail feathers. Everyone either laughed or was in shock. Frankly, I did not care what anyone thought, but I always stayed on the edge of decency out of respect for my online friends. Playing the roll of a female hawk was not in anyway easy, but it was fun.

In my chat room, my friends and I played out our username characters. I could actually be a hawk, almost like a cartoon character, flying or soaring or just sitting and watching. Sometimes it was comical, sometimes sad. You can be anything you want or need to be and that is the beauty of it all. Sometimes you can just be yourself, but that was always boring as hell for me so I tried to avoid being me at all times.

That night I was playing all the sexy tunes I had in my sounds folder, strutting myself across the screen as my hands typed in a frenzy. Flapping my wings to the beat of the music, I pulled feathers out, throwing them at the men in my homeroom. I was playing the part of a real ladyhawk. Everyone laughed and played along with me. Little did they know that I was drunk and in pain. I was really showing my bird butt when Kate came into the room and saw me. Within ten seconds, Kate was sending private messages wanting to know, "What the hell is up!" I ignored the PMs, something Kate can't stand. It took her about three minutes to get a call through before my telephone rang.

"Okay," she said. "What is up?"

I began to explain that I was drinking and was in a go to hell mood. I poured my heart out about Ronnie. Kate and I had been friends since I came into this particular chat room. We formed an almost instant friendship that carries on to this day. We had never met in person but had spent endless hours chatting on and offline many nights. Normally she would have bitched me out about my actions, but she listened to me raging, as only a good friend would have done at that time of night.

She was a great friend and we finished talking three hours later with me ready to pass out. She knew what she was doing in keeping me out of main chat and keeping me from embarrassing myself any further in front of the regulars of our room. Now that is a true friend. When we finally did stop talking, I went to bed feeling as if I needed to tie a rope around myself to keep from falling off my waterbed. Yet the next morning, I woke up with no hangover and was ready to get back online under another username to play. Only I did not have much time to play with my friends today due to the funeral I had to attend in Memphis the next day. My car needed to be serviced before I made the trip, clothes needed to be washed and ironed.

Getting up the next morning, I headed to Memphis, which was about an hour and a half from me. Driving to the address I had been given, the trip took me two hours and several stops for directions. Since I no longer lived in Memphis, I was glad I had given myself the extra time for getting lost. At least I was smart in some ways.

A military funeral has to be the saddest thing one can go through for a friend. I knew I would cry no matter how hard I tried not to. This day would be difficult for me. Yet, I felt like I was ready to be strong for Ronnie. Watching the streets while following the written directions, I finally located the cemetery. One thing about cemeteries, they are not that hard to find when you get close enough. I could see cars in a line behind the hearse and I knew this was the right place at last.

Walking towards the starched and neatly pressed military uniforms, my stomach ached from lack of food and from nerves. I could not avoid noticing the family chairs and the lack of family present in them. I sat at the end of the first row respectfully. Dressed in full military uniforms, four men stood on each side of my friend's coffin. Three more stood a distance off with rifles ready for firing on command. I never heard a word of the eulogy the priest said. His tone was low and my mind drifted to a peaceful place as I watched the leaves in the trees sway in the breeze. Funny I even noticed the many colors that were around me. There was an overpowering sense that Ronnie was saying it was finally okay and thanked me for coming. I was glad I had come to say my last good-bye to him and be happy for his new journey. The uniformed men solemnly lifted the flag in unison from the top of the coffin and folded it into a three-corner tuck. The man in uniform closest to me handed me the folded flag, bowing before me as he spoke words of comfort. He quietly asked if I would like to bid my friend adieu. Nodding yes, I stood up and pulled a piece of paper from my handbag. I began to read in a shaky voice:

WHEREVER YOU ARE

For not looking back when your country called
For going ahead when others were appalled

For being there when muster sounded
For not quitting when your heart pounded

For sweat and sorrow and muscle cramps
For all the fear and pain of boot camp

For every question that wasn't answered
For prayers you thought weren't heard

For leaving your loved ones all alone
For the faraway places, some unknown

For being the one who was so apt
For standing so tall in the gap

For holding true to the dream
For giving your all to the team

For not surrendering to the enemy
For not forsaking the many

For your sacrifice you see
For we are now free

For there wouldn't even be an Uncle Sam
For if not for you sir and ma'am

For your answering God's call
For your unselfish love of us all

Thank you, Ronnie. Fly with the Angels.

    I clutched the flag to my breast as the guns fired three times. I turned to see the men with tears running down their faces. It was time for me to leave and walking slowly back to my car I thought, today will be a day I will never forget. I would always treasure the flag that I held to my breast.

    I kept my mind focused on driving home and things I could do online when I arrived. Trying my best to not think of Ronnie, I let my thoughts run toward my computer. When I was away from my computer, I always felt like there was something missing in my life. Now in my sadness, I needed my mechanical friend. Keeping my mind focused on doing e-mail and surfing, the trip back home did not seem to take as long. The weather turned nasty as I pulled into my driveway, but I was safely home and that was all that mattered. I turned my computer on, made coffee, then changed clothes before getting settled in front of the monitor. If the thunder and lightning crashing all around me moved any closer, there was a chance that I would have to turn the computer off. I checked my e-mail in a hurry, deleted all the junk mail, but still had nothing from Larry. There was no reasoning to this madness and my heart sank again.

In the mailbox was a nice e-mail card from Jim reminding me that California was waiting for me. Listed on the card were dates that would be best for him and as I sat there, the idea appealed to me far too much. Composing a letter back to him, I said yes, that I would be ready a week from today; please send my ticket. There was no sense in wasting a perfectly good trip to California while waiting for an e-mail that would never come. I could not justify sitting at home waiting and driving myself insane trying to find out if Ed was in fact Larry. Perhaps I needed this trip to get far away and see if I could start again. Or at least have a good time. I was tired of living in dreams and on hopes. All my tomorrows were running into each other and looking the same. Whistling California here I come, I made myself smile while surfing online to maps of the area where Jim lived.

Never having been in a plane before or to California, I was very excited. I had my hair done, which was overdue, while I was out shopping for new clothes. Jim was a big shot in his company and I fully intended to make him proud to be with me. My feelings of being inadequate were feelings I had dealt with all my life, but I knew I had enough social training that I could hold my own in any type crowd. I did not like crowds but that did not mean I couldn't deal with them. I would play another part for Jim and the part would be me as an outgoing person, open and free. I was good at fantasy and playing parts on the Internet had helped me become apt.

Sending an e-mail, I asked Jim to meet me online later that night when he had time. We could talk further about my trip. Was I afraid? I asked myself, then quickly replied, not for a single minute. Full speed ahead.

Returning several e-mail messages from friends, I was off surfing the Internet again, looking for anything that caught my attention or fancy.

I had not been surfing the chat rooms long when Ed sent a PM asking if I would like to play. He had also added me to his friends list, which made it easy for me to find him and for him to find me. If he wanted to play, I thought, squaring my shoulders, I was his girl and it would be easy with him since I knew he was Larry, or at least I thought he was. All I had to do was close my eyes and visualize Larry as the one I was speaking with. I sent a reply to his PM saying, "I will play, if you know how to stroke me and make me cum." God, I was bad at times, I thought to myself as I laughed. Oh yes, he typed back, he would make me cum. Let the games begin.

Honestly, I cared nothing about acting this way, feeling it was cheap and degrading at times but it seemed to be what he wanted. At this point, all I cared about was being close to him—no matter what form it was in. Ed and I must have talked for two hours the first time, and another two hours later on that night. We had a good time laughing and cutting up. The sexual talk was funny and I always ended up laughing, sometimes even surprised at the things we both would say. He was very easy to talk to and there was not a conversation between us that did not remind me more of Larry. Each message was getting sexier and wilder than before and I was having a great time. All my Leo traits for acting came alive again, and I felt I was blooming or maybe that I was a blooming idiot, whichever, it made no difference at that time. I was laughing out of pure happiness for a change. This bantering back and forth was nothing like Larry, and surprised me for being so sexual. Larry had always seemed very rigid about passion. I did recall one conversation where he told me that he knew how to satisfy a woman in many different ways. Still, there were never any detailed conversations about sex between us.

During my talks with Ed, I wanted many times to say, "The jig is up, bud, I know who you really are," and yet the truth was, I was never sure enough to even try to say that. Totally confused about who was who, if it really *was* Larry, I wanted to see how far he would go with this charade. One thing for damn sure, he was not a player, and he knew nothing about the art of cybersex. With each PM, I laughed harder at him for trying to be sexy. His descriptive sexual statements turned me on but only because I was playing him hard and only because I loved him so much still. In my mind, I was talking to Larry. Whatever Larry and I did, there was no remorse on my part. What I did or said was out of pure love for him, but playing this game meant that I had to call him by his new username, which was ed_the_bandit, Ed for short.

Sometimes our PMs were hot and steamy but there was never a time that I did not end up laughing. He wanted to play like we were at a truck stop and I was only wearing a long gown with nothing underneath. I smiled and watched as he became very amorous.

Ed: "Honey, I'm holding you closer to me, pulling your dress up and stroking your bare bottom."

LadyHawk: "Yes, darling, stoke me. I like the way your hands feel on my ass. Do it some more."

Ed: "Oh yes I will. I will kiss your neck. Would you like that? How about I spank your bottom some, too?"

LadyHawk: "If you hurt me, I'll run. I do not do pain and I'm certainly not into S&M. If that is what you have in mind, let us forget it right now."

Ed: "No, honey, I will not hurt you, I promise. I will only give you what you want of this dick of mine."

LadyHawk: "So how much would that be? I like my breasts played with, then sucked on."

Ed: "Oh I have a full seven inches honey and it is almost two inches around. Would that be enough?"

LadyHawk: "It will be enough if you can use it well, dear. Tell me more, turn my world upside down if you can."

Laughing at his crude attempts to be sexy, we played on. I shook my head at my monitor in amazement that this man had no true passion. Fantasy was a wonderful thing in the mind but like everything else in life, if you indulge too much, you end up sick of it. Taking control of the situation, I led with all the sexy dialogue. His attempts had me rolling in the floor and horny as hell just thinking about the possibility of this guy being Larry. It didn't take long before I was bored with him and made up an excuse to get the hell out of this conversation. He started out real good, but I was doing all the typing and all the playing, which was turning us both on but was wearing my poor fingers to the bone. He wanted to do more in the truck stops with other men watching. This was a new type of fantasy for me and I needed time to think about it. In the meantime, I had another PM from someone I knew could turn me on.

For some reason, even thinking he was Larry did not bother me any longer. Was the thrill gone? Maybe the fire had died inside or I had died inside after all this time. There was the possibility that I was just tired of beating my head against a brick wall, too. I was not sure what was going on with me. I just felt that this "instant friendship" would not last much longer. Why did the jerk come back into my life now? Was it only to hurt me more? My judgement had been wrong so many times I never counted much on it now. If this man did not turn out to be Larry, then I had no reason to continue because Ed was boring as hell.

Saying, "I have to run, we'll talk later," I made away like a jet plane to my waiting friend. At this point, as much as I loved Larry, I did not want to be his toy. My self-protection mode kicked in and it was on standby. Without a doubt, I was losing it again like I always did with him and that is what I hated about myself so much—no control with him.

Chapter 7

Voice chat was now the big thing to do. Yahoo had made it very easy and I loved Yahoo for giving us voice. Putting my headset on, Bobi and I begin to play. He was another friend of mine in chat who always flirted with the ladies and who I thought was so handsome after seeing his picture on his profile. In voice, he was telling me of things he wanted to do to my body, where he was touching me. I talked of stroking his back with my long nails and grabbing his ass, pulling him closer into me. I could hear his breathing become faster as we talked and I was getting hotter with each passing second. He was telling me how he was kissing my neck and ears and making his way down my stomach, and I spoke of returning his kisses and licking his nipples, feeling them growing harder in my hot mouth.

"Oh my, yes," he moaned, "give me more, honey, give me more."

Whispering softly, I told him I was moving down his body continuing to lick and kiss on him. New batteries were in my vibrator and as we spoke, I used it masterfully on myself. With my extra headset line, I could lie on my couch and actually get into the fantasy of having him with me or on me. The more he whispered in my ear, the more aroused we both became. I knew that moaning softly was a great turn on for this man and I would give it to him. It took about five minutes for us to come to a full climax, which I must admit was very rewarding for me. From hearing his voice over my headset, I knew he had taken care of himself, too. Laughing together about how good it felt, we talked a few more minutes about other things in chat.

I reached for a cigarette and a drink of my ice tea trying to come back to earth and settle down. This was a wonderful way to begin my day and I told him so. We talked a few more minutes and said our farewells until the next time and yes, there would be many times for us ahead. There would be no falling in love or commitments between us and it was another ideal relationship for the both of us. I loved life at these precious moments. Except for my pain, I thought I was doing fine. Taking more pain pills, I was ready again for anything.

Within the next month, I was in the hospital having my back surgery. The pain had finally become too much to bear. No amount of pain pills seemed to help. This would delay my California trip for about six weeks, but I would be out of pain for the first time in many months. After the surgery was over, my doctor came into my room and explained what he had done and that I would also be able to come off the pain pills now. In fact, he was going to send a specialist in the area of pain to come in to talk to me about that. The main thing, I thought, was no more pain like before. Now I needed a smoke and to get back to my computer. I was very happy that I had gotten this over with.

Later that day, I was moving carefully around my room when a very nice lady came in to talk with me about my medicines. She asked me how I felt and I told her great, then she said, "I see from your chart you just had 100mgs of Demerol, IM," she said.

"Yes," I said, "about twenty minutes ago." Her next question was to ask how long I had been taking pain medicines. Counting back, I told her almost eight months. Sitting on the edge of my bed she told me that my body was now addicted to pain medicines without even blinking an eye. She asked if I had experienced any mood changes or visions and I told her mood swings, yes, very intense mood swings, but no visions. She talked about how pain medicines react with your body and also your mind. When she finally left, I was amazed, not to mention in shock. She told me that it would take months for all of these drugs to get out of my system and even longer to get over the dependency. Up until that time, I had not thought about it but I knew she was right. She told me the best way to handle the addiction when I got home would be to come off my pills slowly and to regain normal activity at a pace good for me, and that is exactly what I did.

It took about three weeks to get myself off the pain pills and I felt wonderful, but I still had a long way to go before I knew I was free. The hot flashes were terrible and sometimes I would get the shakes so bad that I couldn't type. Jim was always there on the phone giving me support, being my rock during the whole thing. There was no doubt in my mind that the man loved me, but I had to get back to some kind of normalcy before I could do anything about it. No one ever said the road would be easy, but it was one I had to travel alone. At least my mood swings seemed to get better with each day, but I was trying to be more careful now. I found myself thinking about things before jumping into them and my mind was getting clearer each day that passed. I was happy and content to some degree with myself. Looking back over the past months, I could see many mistakes that I had made because of my irrational behavior, but there was nothing I could do about all that now.

I had turned into a new woman. I was the LadyHawk again, and nothing would stop me now from living. Each day was getting clearer for me and my head was not as foggy as it once was. No need to lie and say that everything was totally grand, I still had times of deep depression and confusion, just not as often as before. I was still uncertain as to what I wanted to do about everything, but Rome was not built in a day. As for now, I knew my house needed cleaning badly.

While doing my chores around the house, I played music and danced around my small apartment as I cleaned. Occasionally, I would glance at my computer checking for any additional messages waiting for me. My screen saver was set so I could go outside for a few minutes to play with the dog and not miss anything. Working quickly, I finished my chores and was sitting smoking a cigarette when Jim's PM came asking if I was having a good day. "Yes," I replied, "I am having a great day and even better now that you are here with me."

He said that he could not stop thinking about me and how much he adored my pictures. Also, how my slow southern voice was so exciting for him that he could not help but to stay horny all the time. Smiling as he spoke, I knew the tickets were in the mail. California here I come.

Later that night, Jim called and as we talked, I surfed the Net quietly searching the pages for any information on California again. I felt as if I was riding on air thinking happily of being with Jim. Being the perfect gentleman, Jim told me his home and computer would be available for me while he was away working. However, when he came home, I would be all his. His suggestion sounded like a winner to me and within a week, I would be flying for my first time. I, LadyHawk, would actually take flight, I thought, smiling to myself. Closing my eyes, I could picture myself soaring in the open sky like the brave bird I pretended to be.

The next week was terrible for me. My hair needed trimming again if I could find the time. Looking through all my clothes, I found that most were not suitable for my trip, but having a limited income, I would have to make them work. Jim knew my financial status and said that he understood. Thinking back over our conversations, he talked many times of his work in the maintenance department of a local factory. He spoke often of his ex-wife, "taking him to the cleaners" and how he had tolerated it because she was disabled. According to him, he did not love her anymore because she and her children had only used him like a dirty sock. His voice would become strained as he spoke of her. I never pushed him into revealing any information concerning his divorce or about his work and speaking about it only agitated him more. Other than that, he was a happy-go-lucky type of guy. When I met him on the Net, his name was "happyman2010."

Talking to friends online about my upcoming trip, they were happy for me, offering congratulations, friendly warnings about being careful, and giving advice on where and what to see on my visit. Being far too excited to stay in one place long, I left my chat buddies to surf the Net. I would return later to check my e-mail.

Out of the blue, Ed's PM arrived on my screen. I sometimes wondered why I bothered talking with him. There were no words from him when I needed him but still, something pulled at me to hang in there. All the evidence pointed to Ed actually being Larry more and more and seeing Larry's name come online but not speaking to me had shaken my self-confidence to the very core of my soul. Yet, the thrill of seeing his name light up was becoming easier for me to deal with. I could not make him love me; therefore, I dealt with that and went on with my life. Ed, however, was a different ball game. He talked of sexual fantasies that I never thought Larry knew about. It almost seemed as if Larry and I were soulmates because we were always being pulled together for unknown reasons. Too bad he just couldn't see that.

During one of our fantasies, Ed invited me to tie him up and use a strap to hit him lightly as we worked out our game that day. "What a nut case," I thought as another question immediately entered my head—was Larry testing me? Whatever the logic was behind his madness only he knew, but we ended up talking for hours about truck stops, convertibles, and white gowns. Whatever he wanted I knew I could provide. A true player online can always deal with the unexpected and dream when called upon to do so. This day would turn out to be very interesting as we began to play our game. This would be a new fantasy for him as I plotted out the story in my head.

Oh, how I loved to play and being with Ed was providing me with an abundance of material to play with. At one point, I looked up at the screen and he had typed in two words, "Love you." I never replied to that. I just continued to type and play. Although, I thought it was a wee bit nutty, he seemed to get a larger boner playing our game that day. Chuckling at the day's theme, we went from straight to kinky in a few short lines. Promising today's adventure would be different, Ed continued asking in PMs what I was wearing. Feeling very confident, I took control.

"Let's go for a ride in your car," I said to him, and we were driving down a twisting highway. I was dressed in a long, white, flowing gown with the shoulders falling down. I would have no undergarments on and my hair would be flowing in the wind as we drove. Of course, we would be in an open convertible. Ed seemed to have a thing about truck drivers and voyeurism. Showing my breasts as we drove by speeding eighteen wheel trucks was strange but highly erotic, too. I started typing rapidly, wanting to make him breathless.

"As we drive down the road very fast, passing everything, I put my knees into the seat, turning toward the passed trucks letting the wind carry the light material around my body up and down. My hair is everywhere but who cared as my breasts are exposed to one truck drivers delight, not to mention my bare ass showing. The truck driver sounds his horns as if giving his full approval of what he is viewing and as he pulls the cable sounding the horns, I hold up my wineglass, saluting him with a knowing grin.

Driving on carefully, you glance my way with a smile in your eyes. Reaching for the radio, I turned the music up loud as we pass another truck. Perching on my knees in the car seat, I begin to rub my own breast as I close my eyes, getting lost in the music. My body responds as I dance in the wind. Without a thought, I pull myself up in the seat becoming a Goddess, demanding the attention of the universe as the music becomes louder. Not missing a beat, my body sways in the wind that holds me in place as we drive. When the song ends, I slip into my seat glancing your way seeing the huge bulge in your jeans knowing that you are excited."

Looking up at my screen and seeing nothing from Ed, I type in, "Baby???" As soon as I hit the "enter" button he responded. "Please don't stop," he said, so I continued typing.

"Please pull over at the rest stop down the road," I demanded. "I'm all hot and bothered. When you pull in, park at the side so all the truck drivers can see when they pass us."

Looking my way, smiling, you ask, "What now?" Reaching for a sack under the front seat, I pulled out a pair of handcuffs and my handgun. Your eyes widen in surprise, but there is no need in worrying. Sliding your hands behind the seat, I slap the handcuffs on, rendering you helpless. You cannot escape.

"Hey, what are you going to do with that gun?" you ask me.

"Nothing," I say, "if you obey." I watch your face become afraid. Laying the gun on the dashboard, I rip your shirt open, straddling your lap with my body. Turning the car radio up loud, I move my naked figure over the lump in your jeans. Slowly grinding my hips over your jeans, I hear you moan. Thrusting my hips into your hardness, I feather kisses on your face and neck, biting your earlobes. My mouth and tongue take possession of your nipples, sucking them hard as you gasp.

"Please, undo the cuffs," you ask.

I shake my head and a no silently comes from my lips. Smiling, I undo your pants, leaving you completely exposed as the truck drivers pass by blowing their horns in approval. You ask me to stop but there is no stopping now. I grab the gun, hold it to your head, and tell you to shut up and do as I say then I lay it back on the dashboard for you to see. Positioning my body over you, my long, bare legs on either side, I slowly lower my body down onto your lap. The steering wheel braces my back. My hands seek and rescue the only thing not exposed and I slowly begin kissing your face, your neck. Easing my own body down further, your hard dick fills me. Unable to move, you lay your head back against the headrest. Delightful little moaning sounds come from your lips while I ride your lap like a child on a merry go round pony. I feel you becoming more intense. I bend down and whisper in your ear, "Are you ready, dear?" A smile comes from your lips and I move faster. Rocking on you until I feel you begin to release, I also let go. Shaking, trembling, and out of breath I collapse against you and you whisper to me, "Look at the truck drivers, honey." Glancing at the truck drivers, we see them waving and hear them whistling their approval of our actions. We look at each other and laugh. After I remove the handcuffs, you start the car then begin to drive, pulling onto the main road. I grab the gun and return to my place beside you. Still looking at me and smiling, you ask what I intend to do with the gun. Smiling as I point the small pistol at your face, seeing your eyes grow wide with alarm, I say, "Darling, I will always love you," as I squeeze the trigger of the gun slowly. The look on your face was worth it all as I slowly dry the dripping water from your chin and smile at you.

"Wow!" he said, coming back from my story on the screen again. "Good lord, woman, you're nuts but great and that is all I can say." Laughing and leaning back into the chair I thought to myself, you have not seen anything yet, my friend. Backing off, I told Ed I had get some things done around home here, to take care for now. I signed off with "X's & O's." (That meant kisses and hugs.)

I shut down my chat screen, satisfied with what had just happened. Could I be as passionate in real life as all the sensual words I typed to my lovers in chat? I was not sure, but I was damn well going to try before long. Jim would be the one I would find out with and I closed the computer down to give it a rest so I could reflect and clean house.

I was really into cleaning my cave when the telephone rang and Jim's sweet voice was there. Something about his voice made me forget everything else I was doing. He asked me how I was doing, which was his style. He was always concerned with my well being and that is what drew me to him I think. Most guys I knew just wanted to cyber, but not my Jim. I had tried to play with him but he would stop me saying I was his lady, and that there would be time for that when we met. I was not afraid he would not be a good lover. Many times when I spoke softly to him, I could here his breathing becoming harder so I had no fears about his being able to make love. Jim said I would have my tickets arriving at my door before the day was over and I was very excited and all packed, too. The next morning I would be flying the friendly skies heading West. I would be all set to go after attending a few last minute details online and making a few telephone calls to close friends. I decided not to say anything to Ed about my trip. Since he claimed to be a player, I would treat him like one. I would let him miss me, thinking at the time that it would do him good.

Maybe if Larry/Ed thought he was losing me it would open his eyes, but maybe not. My thoughts of Larry were getting far and few between and only when I played with Ed did I really think of him anymore. Occasionally I would send pretty web pictures to his old address. Sometimes, getting in a lonely mood, I would write very sad letters to him saying how much I missed him but there was never a response. I was trying to act cold hearted toward him but I would not burn my bridges just yet. The game was getting good.

Saying bye to my best friends online that night left me pumped full of new, excited energy. I felt like the future was open to me but it's hard sometimes to forget the real heart yearnings.

My youngest son drove me to the airport the next morning giving me a hard talking to about his being worried. Oh my, I thought, here I was getting a lecture from an eighteen-year-old but it was good to know he was concerned about me, too. We arrived at the airport and I have to admit, I was shaking inside. For some reason, the planes had always looked bigger on television. After hugging my son, I boarded the plane and took a seat next to the window with my camera ready. I wanted pictures of everything and took many of the sky and cloud formations as we flew. The trip lasted four-and-a-half hours and, feeling silly, I had to go pee at 35,000 feet. This country lady was experiencing many firsts, I thought, as I smiled to myself. I must have stretched my neck two extra inches trying to look at everything from the window.

It seemed like we had just gotten off the ground when the pilot announced that we would be arriving in San Francisco on time in about ten minutes. Not being familiar with all the airport protocol like baggage claims and several other details, I began reading the ticket information again. I think the normal fears of losing my luggage and getting lost were raging in my mind. I laid my head back to do some deep breathing. That did help but then I felt the wheels touch ground and my heart raced again. Oh my goodness, I was in a strange land for sure now and said a small prayer to let me get through this and to protect me.

I started following everyone off the plane, afraid I would get lost while juggling the small cases I had carried on. As I stepped into the terminal, I saw a man holding a sign that said LadyHawk.

"I am LadyHawk," I said, walking up to him, but this was not Jim.

"Mr. Martin asked me to pick you up and deliver you home and said to say how sorry he was, that something came up at work that needed his immediate attention."

Oh my God, I thought, but the only word that came out was, "Okay."

He took my bags and baggage tickets and I followed him all the way outside the building wondering what was next. Trying not to lose sight of the driver, I followed closely behind until I reached the curb where the driver helped open the door of a long, white limo. Feeling my knees go weak from the shock, I was feeling faint, but my legs were killing me from the walk.

Looking at the man with the sign, I asked in a bewildered voice, "Are you sure you have the right lady?"

"Yes, Miss, I do, I have the right lady... Here, see the picture."

Glancing his way, I saw the picture I had sent to Jim several weeks ago. Yes, it was me alright. My mind was in awe as the man loaded my baggage into the trunk, then started driving.

"Is there anything you need, Miss?" the driver asked in his polite voice.

"No, thank you," I answered and settled back into the seat wondering what was coming next. Smiling I kept thinking, a limo.

After driving several miles, we were approaching a suburb. The houses set back off the main roads and each were what I considered mansions. I had pictured Jim living in a trailer with a few dogs outside, but this was far different than I imagined. Suddenly, we were pulling up to this large, gated entrance and the driver slid a card through the slot in the mailbox-looking pedestal and the gate opened. I was having a panic attack on the inside, but whatever was happening I knew I had nowhere to run so I did my best to remain cool. After driving down the tree-lined drive, we pulled up to the front of the grandest house I had ever seen. Jim would have lots of explaining to do about this.

The front door opened and a little Mexican lady in a maid's uniform greeted me warmly. She welcomed me, telling me to follow her in please. The driver followed, bringing my luggage from the trunk of the car into the hall and then disappeared. The maid asked that I call her Linda and explained that she was taking me to my room and that Mr. Martin would call soon. Opening another door, Linda beckoned me into the largest bedroom I had ever seen. My whole apartment could fit in here twice, I thought, as I looked around. Linda said she would bring up a snack and I should make myself comfortable.

After she closed the door behind her, I stood in the middle of the room thinking myself an idiot and not knowing what to do next. Exploring my surroundings, I found a bathroom that was made completely of marble with a whirlpool! This had to be a joke, I thought, then asked myself, What the hell am I doing here? I was laughing all the time.

I heard a telephone ringing but for the life of me I could not find it and it soon stopped ringing. I sat on the king size bed bouncing like a child when Linda came back into my room with a tray and a cellular telephone. Setting the tray down, she smiled and said that Mr. Martin was on the line.

"Jim," I almost shouted into the receiver, "what is happening here, is this a joke?" He laughed saying, "No darling, no joke. You are in my extra bedroom."

"Darn, Jim, you did not tell me you were rich." He laughed and so did I. "So when is "Mr. Martin" coming home so I can personally thank him?"

"I will be there in about two hours and in the mean time, Lady, why not go outside and enjoy the pool," he said, laughing at me. "Linda has instructions to aid you in whatever you want and if you will look in the closet, I have arranged some extra party clothes and a bathing suit for you," Jim said. "See you soon, darling."

I was walking around the room trying to find the closet when Linda appeared again asking me if I needed anything else. Looking at her, puzzled, I asked, "Where is the closet?" We both laughed as she pulled the mirrors back revealing the closet doors behind. Oh my, I thought, looking through the clothes and bathing suits Jim had placed there for my use. This meant there had to be a pool so I walked over to the window looking down to the courtyard below and saw a large, oversized pool surrounded by plants and trees making it appear tropical. Okay, I knew I was dead and this was heaven. Laughing, I found a suit that would fit and headed out of my bay doors to the pool. I could not wait to tell the girls about this and I forgot to bring my camera along. There was still time to take pictures though; I had a week of this grand lifestyle to deal with. Yeah, I could handle this.

The water was great and I played like a child turning circles underwater, living the fantasies I had about being a dolphin. This was so wonderful. Being too old to act that way for very long, I drifted over on my back and floated until I heard a voice.

"Jim," I screamed and laughed as I jumped out of the water and started hugging him. "Why did you not tell me about all this?" I asked.

He said he did not want me to care for him because of his money. It was not my style to place money above people. He wanted to take me on a tour of the house, so grabbing my robe and his hand, off we went.

Looking at Jim as he talked about the house, I wondered what his wife had walked away with after seeing the priceless treasures in every room, but it was none of my business. I was just happy to be there. Jim seemed like a "good ole country boy" dressed in his jeans and a western shirt. He did not look as if he belonged here anymore than I did. I moved in front of him and kissed his mouth passionately. When I pulled back, he had a very nice smile on his face.

Looking at me, he said, "Do you love me now since you see I have money?"

Taking a breath, I looked at him and said, "Jim, in my heart you will always be my friend and God has blessed you with many fine things, but let's give it time and see where it goes. You know, I am not in a hurry," I said, looking away from his eyes.

"Your only problem, Veronica, is that you're still in love with Larry," he said. "You can lie to everyone but you can't lie to me because I know your soul, my friend."

Grinning at him, I asked to see that computer room I'd heard so much about. Yes, I was changing the subject and he knew it, as he smiled that beautiful smile at me again.

"Holy moly!" I gasped as he opened the computer room door. Looking at him, I said, "I didn't know NASA worked out of your home." We laughed as I examined every monitor and keyboard. He had the largest scanner I had ever seen—this man had it all. All this was too good to be true. Jim told me to play with the equipment if I wanted to while he changed for dinner.

Maybe later I would play on the computer, I thought, right now, I needed to take a bubble bath and soak awhile, maybe relax a bit more. Jim and I walked up the stairs pausing as he opened my door and kissed me on the check. Flashing him a smile, I said, "Later," giving him a wink. He replied, "Most definitely," as he kissed my hand. Looking down at my feet, I smiled. Yes, they are still touching the floor, I thought, laughing.

Inside my room, I found the invisible closet again and just as I thought, there were two suits and three nightgowns just my size with shoes to match. "Damn, this man is wonderful and you are one stupid bitch," I said to myself. Grabbing the robe, I headed into the bath feeling like Julia Roberts in the movie *Pretty Woman*. I could not stop laughing and giggling.

I must have drifted off because a loud knock on the door made me jump straight up in the water.

"Honey?" Jim called.

"Yes, I will be right out. Give me about ten minutes," I shouted to him outside my door. Hearing his chuckle, he mumbled something about women and I made a mad dash for my dinner clothes. The suit was light gray and the blouse was white silk with a matching jacket, of course. The skirt was straight with four-inch slits up each side. Boy this man knew his clothes and me, I thought as I opened my case and pulled out the up lift bra and a black garter belt for the black, silk hose. Reaching for my high heels, I nearly fell over backwards. I saw myself in the mirror and I turned around to see who was in the room with me, but no one was there!

My ten minutes were up as I run out the door and headed down the stairs. Jim was smoking a cigarette, pacing like an old mother hen. He looked up at me and the most amorous smile spread across his face. My weight was still way overboard but I knew I looked grand. I came down the stairs slowly, quite unsteady from wearing those high heels, but it was for Jim so it was okay.

At the bottom of the stairs, he held out his hand, taking mine and kissing the palm softly. He then wrapped it around his arm. Smiling, he talked about how wonderful I looked and how glad he was I'd come out for a visit. Leading me into another room, he walked me over to a table filled with the most gorgeous display of food I have ever seen. Linda was pouring the wine, then lighting the candles before she turned to leave. Jim held the chair for me and we laughed and continued talking, playing like children with our food and drinking wine like it was water. Music filled the background coming from some mysterious place in the walls and my head was beginning to spin. I excused myself for the powder room, trying to maintain my balance in my heels as I walked. I could not help giggling at myself. I splashed water on my face trying to get my eyes to refocus, then redid my makeup. Glancing back into the mirror, talking to my on image, I said, "You are a lucky lady so you go back out there and make Jim a happy man." I winked at my reflection in the mirror.

Walking back to Jim I noticed that the room had gotten much warmer and the music much louder. Jim took me in his arms and we started to dance. The temperature in the room was too warm and I had to come out of the shoes and jacket. Maybe it was the fire in the fireplace or the wine or a combination of them all, but I was burning up. Jim's body dancing so close to mine felt like a furnace. He pulled away for a second talking to me, then saw my face.

"My God, darling, are you okay?" he asked.

"Yes, just very warm," I replied. Jim laughed and said it was the wine. He grabbed my hand and pulled me toward the patio doors. Grinning, he asked if the pool would be better for me and I laughed saying I did not think I could make it upstairs for my suit. He grinned then and suddenly began pulling off his clothes and betting me that he could beat me to the pool. That was all I needed to hear. I was coming out of my clothes, but Jim beat me. Splashing around in the water like a kid, I watched and laughed with him, finally getting butt naked, too.

I looked at him. "Where is that music coming from?" I asked.

"It is all for you, dear," he said, holding out his hand as I walked down into the water. We played for about an hour when Jim grabbed my waist—about to dunk me—then he stopped and brought me down in front of him. Up until this moment, I had not actually seen anything of his manhood in the water, but now I could feel it. Sliding my arms around his neck, we kissed. We kissed many times as his strong hands held and caressed me. God could the man kiss. Finally, he pulled away, pulling his body up on a water raft. I smiled as he held his hand out to me. Pulling me up from the bottom of the raft, my body slid up on to his. How he did what he did next just blew my mind. The music got a bit louder and the lights around the pool starting blinking with the music, changing colors, and dancing sprays of water encircled us. What a wonderful show this was. Jim was marvelous. Totally amazed, I leaned my body down onto Jim's, watching and listening for a second, then I moved. Sitting straight up over him, I placed his hands on my breasts and gently rocked my naked bottom against his. I could feel his hardness pressing more firmly against my body.

"Lord," I said to myself with my eyes closed, "if I am going to die anytime soon, please let it be now." I must have shifted too much because into the water I fell, with Jim following. Laughing hard, we left the pool, deciding it was time to call it a night. Actually, Jim decided, but I followed. Reaching the stairs, Jim said for me to go dry off and he would come in to kiss me good night.

Okay, I thought, did this mean there would be no sex between us? I was not sure but I knew I would not push him. I headed to my room to dry off and to get ready for bed. Pulling down the huge bed covers, I found a beautiful, long, strapless nightgown. Dancing around the room holding it to my breast, I let it fall softly, covering my frame. I stood admiring the image in the mirror and I heard music again. I must have smiled more this day than I had in ages and now my favorite song was playing, Leon Russell's *Stranger in a Strange Land*, which was true for me. The lights went dim when my door opened and Jim was holding one single rose looking like a Knight, so tall and handsome. This was too much for this country girl to handle. I took the rose, smiling at him, then taking his other hand, I led him to the bed. Lying with arms entwined, we talked a while and drifted off asleep.

Sometime later, I was awakened by gentle hands caressing my body—sweet kisses on my face and neck. His warm lips touched my body while his fingers traced my legs, running up and down. I shivered with excitement. Not fully awake, my arms were reaching out to touch back but he stopped them, gently placing them above my head. With my eyes still closed, I could not help but move as I felt Jim's mouth engulf my nipples, sucking them like a child nursing. Wow, I thought, what a way to be woken up in the morning. With ever so much ease, he opened my legs, rubbing his fingers through my hairs. Little tingles raced along, over my skin as I moaned in my quiescent state. Jim leisurely placed himself on me. Now fully awake, I moved only to help him reach his goal. His hard dick worked inside of me and every time I moved to touch him, my hands were held back. As he thrusted in and out of my body in a steady rhythm, I knew I was about to cum and my body responded accordingly. I could feel the throbbing starting to decrease along with Jim's breathing and, wrapping my arms around his body, we drifted back to sleep with him still in me.

Later, I opened my eyes to find Jim gone and I could hear rain beating against the windows. My first thought was that I needed to get to the computer room and check my e-mail. Even as wonderful as things were, I did miss my computer and my friends. Should I dare hope that there would be an e-mail from Larry. It was always Larry in the back of my mind. Where there's life there is always hope, were my thoughts.

Jumping into my clothes, I ran downstairs to the NASA room, as I now called it. Getting all that "high tech" equipment to work may not be easy for me, but looking at all the buttons, I pushed the one that said "ON." That turned out to be easy enough and I was on the Net surfing again. Linda brought in coffee with toast and I continued to play online the rest of the day.

There was no e-mail from Larry. Yet, I had no sooner entered the chat rooms than I received a PM from Ed. He wanted to know where I had been, almost like a demand, and I giggled. Ignoring his question, I asked how he was and he started to tell me how lonely he had been without me. My thought was, asshole you get what you put out in this life, but I never said it. Maybe things were not as they appeared with him, but I wanted to make it with him now and cybering was the only way I could. Locking the door, I started typing to Ed/Larry about what I was doing to him and what I felt like while I was doing it. God he was easy and fun. Maybe he was learning now the true meaning of cybersex. I was a great teacher and I smiled while I waited for his response. This day, Ed wanted to ride in a boat while we made love and that worked for me as my mind raced ahead of the storyline I was to use for this playtime.

He described the boat and said that we were just riding around the lake near his house and that it was almost dark. Of course, I was naked lying back in the seat with my feet propped on the dash as we bounced over the water going very fast.

"How does one do it in a speedboat?" I asked myself. I let him take the initiative since this was his idea. Ed never wanted the romance, only the kinky stuff like in boats or driving down the road. Hell, he even wanted to make out in truck stops. This time I'd had enough of the bullshit. Typing in my one last message, I said I had to run and next time for him to come up with a better storyline, but I said it as a joke. I sent a PM telling him, I had company, and I would catch him later. Poof, I was gone!

Actually, I did have company, it was my other online lover, Dodger101. Dodger and I had been having an online affair for about two months. He was in Ireland and I enjoyed his accent. He could speak to me and my pants came off. I did not have my little fun toy so we ended up talking about everything else with a promise for more later. Knowing Dodger, I knew that when I got back home he would do lots more than talk. I had e-mails to deal with and then I went surfing, bouncing back into my chat room often to see if anything new was going on.

I spent the rest of the day surfing and never heard Jim knocking on the door. He was trying to get my attention without startling me because dinner was ready as he tapped lightly on the door. Finally I did hear him and I cleared offline, closed down the computer. Laughing at my own simple minded thoughts I ran passed him giggling like a teenager and then upstairs to get dressed. I rushed into my room heading for the bathroom, then did a double take, looking back to my bed where a very big package sat. "Damn, Jim," I mumbled, "you are unbelievable." Folding back the tissue and looking underneath, I saw a beautiful red sweater with a card saying, "You are the greatest, love Jim." Shaking my head, I laid it on the bed and jumped into the shower.

The whole week went that way and it was the same every night, too. Jim always appeared at my door with a flower and I always woke up being made love to. I wasn't complaining, in fact, I actually thought everything was wonderful except I did miss home. I did miss Rebel, too.

The next day, I was at the pool when I saw Jim coming out toward me. This seemed strange 'cause I knew he was supposed to be at work. He sat down in the chair after he kissed me hello and asked we if we could talk.

"Sure, Jim," I said, "talk to me about anything."

Speaking in a strained voice, he stated how much he had always wanted a woman like me and said I could stay with him forever. After making that statement, he handed me a small box and I looked at it, stunned, knowing what was inside. Jim said, "Look at it and think about what it means. I have to make a run into the company and I will see you about six o'clock, darling," then he was gone.

I was still holding the unopened box and I knew it was time to do some serious thinking about all this. It took an hour for me to find the courage to open that little black box and inside was the most magnificent ring I had ever seen. Yet, when I looked at it, my silly heart kept calling Larry's name. Closing the box, I jumped into the water trying to drown my fool self, but I always floated back to the top. Damn, what was I going to do now, I wondered as I floated on my back just looking at the clouds.

My plane would leave the next day for home, but tonight, Jim and I would talk about what was going on. At six o'clock sharp, a knock sounded at my door and calling out, I said I would be there in a second. Jim smoked occasionally when he was nervous, but I had been puffing all day while I was thinking. Taking a deep breath, I headed downstairs. This time Jim looked shocked but said nothing. I had my sweat pants on and a floppy shirt, no make up, and not much of a smile. When I started downstairs, I could actually be sure I was not going to fall because I had my tennis shoes on.

At the bottom, Jim was waiting, but the radiant smile he normally had for me did not appear. With a knowing grin, he took my hand, kissed it, and led me in to dinner. Linda served dinner while music was playing in the background. Candles were casting a glow over the lobster and fruit, and it was wonderful. Now came the hard part and I was not sure how it would go, but my heart controlled my mouth. Setting the ring box on the table, I invited Jim to pull his chair closer to me.

"Jim, I have to say that this has been the greatest and the grandest experience of my life. Nothing will ever top the memories for as long as I shall live. Still, if I cannot come to you in total mind and body, then the relationship would only be a lie. You know, Jim. Besides, I have trusted you like a true friend. I have told you how my life is and how my heart goes. Until I can come to terms with myself and get my head back on straight, I am not going to marry you." Seeing the sadness in his face broke my heart and I couldn't even force a smile.

Lifting my hand again to his lips, he kissed it and said, "Veronica, the spirit you have and the true honesty in your soul is the reason I love you and that love knows no time frame. You go back, work your life and feelings out and we will always be friends. If you ever need me, I will be here for you."

My inner self was screaming at me, "*Fool*!"

Hugging each other tightly, we started grabbing food, eating like crazy and stripping for the pool. This was my last night, so why not make it good? Jim showed off on the diving board while I clapped and whistled. Waiting for him to surface from his dive, panic suddenly swept through me as I started looking underwater for him. He was floating about halfway down. To be honest, I did not know how well I could still swim, but I was diving for him. As I brought him up into the air, he gasped and spit out some water and I continued working to get us to the closet edge of the pool. Luck was with us because he'd only had the wind knocked out of him and after a few minutes, he was back to his old self laughing and playing.

Within minutes, we were holding each other in the water. His big strong hands were rubbing my back and my arms were around his shoulders as I nibbled his neck. My legs locked around his hips as he held me tenderly with so much love I could feel it pouring into my own body. I eased myself backward into the water facing the open sky. Jim placed a float under my back to help keep me up while he rubbed my breasts, maneuvering my body so that it would gently rock at his command. He pulled us into the shallow part of the pool, whispering for me to please lay still and watch the stars. Doing that, I could hear the music playing in the background and could tell it was Bob Segers' *Chances Are*. Jim had set up all my favorite music, being his thoughtful self. He also knew that certain music was like adding wood to my fire. Opening my eyes, I was looking at the heavens above me filled with the brightest stars and floating on air while a wonderful man was sweetly devouring my body. With all this grandeur, my heart still cried for the one I truly loved and I felt so lost.

Jim and I continued to make love very slowly for hours, stimulating each other and bringing each other to one orgasmic moment after the other. The last thing I remember was Jim and I lying in a large lounge chair cuddled up, where we must have fallen asleep.

Smelling fresh brewed coffee teased me into stretching. Opening my eyes, I saw Linda bringing a tray for me. God I wished I could take her home with me; this was great. Smiling at her, I glanced down at my body making sure I was covered with my large towel. She said that Mr. Martin had left a note for me upstairs. Lying there a few seconds longer, sipping my coffee, I glanced at my watch and knew I had to get moving. My flight would be leaving in two hours.

Climbing the stairs to dress and pack, Jim was nowhere to be seen but there was an envelope with my name waiting on the bed. Not opening it, I placed it in my bag and got myself organized. Placing all the wonderful clothes back into the closet, I heard a soft knock at the door and Linda telling me that the driver was ready. With one more look around, I grabbed my bags and closed the door, feeling depressed and very lonely as I left.

During the flight back to Nashville, I opened Jim's letter and smiled. He had written, "I will always love you and when you are ready, I will be here waiting." That is all it said, and I guess that is all that needed saying. This time on the plane I took no pictures, instead I held back the tears and remembered the past week. Making my way into the restroom, I sat there and let the tears roll silently. It was not long before we were landing at the airport once again. My youngest son greeted me with his smile and a big hug. I knew he needed gas money and some pocket change just from looking at him.

"Guess you missed your old mom, huh?" I said. He grinned and said, "Only because I needed to borrow ten bucks!" We both laughed and headed home.

Chapter 8

Arriving back at my very small apartment was great. I had hired a neighbor to care for Rebel while I was gone and he was jumping all over the yard as happy to see me as I was to rub his face, getting tons of dog kisses. Going upstairs, I opened the door and the air was stuffy, but there in the corner just as I had left it was my friend, my computer. I turned it on to go online before I even unpacked. I sent out notices that I was back and safe but there was nothing from Larry. Turning on Yahoo pager, I found six private messages from Ed wanting to know where I was. There were also about ten more messages from my other friends so I began to answer each one, letting them know that LadyHawk was back.

It had not been two minutes before I received a PM from Ed demanding to know what was wrong and why had I not answered him. Sitting back, I wondered if he even deserved an answer. Then I thought of Jim and the reason I'd come back home. No longer addicted to the drugs like I once was, I was still confused when it came to this situation with Larry/Ed. The mystery behind it all was just that, a mystery, and one that might never be solved. I was not burning any bridges now. There was still a slight bit of hope in me that things would one day work themselves out.

"I have missed you, Ed and I am glad to be back," I said, then I told him the truth—well, almost. I had almost forgotten that this was still a game. Ed did not know that I thought he was Larry still. Through Jim's NASA setup, I had tracked down Larry and there was very little doubt in my mind that Larry/Ed were the same person, but I was not one hundred percent sure. Funny thing was, I knew Larry had moved to New York State and this Ed character lived only a hundred miles from Larry's new address according to him. I had Larry's new telephone number and I called it once only to hear his voice say hello before I hung up the telephone. I never called it again. I only wanted to be sure that the information I had was correct, and it was. If Ed wanted to continue the game, then so be it. I was a player and could handle it now. I told him I had been at my mothers for a few days because she was ill and needed me. Whether he believed me or not I did not know or care, but he did accept my answer without anymore questions. We only talked a short time and then I said I would return later for fun and games. He did not seem happy over that and I needed to do something, but was not sure what. I needed more time to think about all this new information with Larry/Ed.

Being a player meant staying cool and collected so I talked to Ed for five minutes longer and then the telephone rang. It was "Rodeoman," another friend from the Internet, except this one seemed "hell bent" on giving me trouble. I had tried on more than one occasion to let him go easy but he still hung on. Apparently, he was not used to rejection and he wanted to talk with me about us. As far as I was concerned, there was no us, but I sat back and listened to him as he jabbered. In the meantime, I told Ed in PM that I would catch him later, that I wanted a nap before I started chatting for the night. Now he wanted to give me attitude, too. Not having much choice in the matter, Ed finally said, "See you later," with one short x & o at the end of the PM before he disappeared. I had to giggle at that—life was that way sometimes.

Gosh, I really was not in the mood to be harassed by anyone. My temper was rising with each second that passed. Rodeoman wanted an explanation as to why I was not talking to him and why I had not been home to receive his calls and that made me extremely angry. I flew into to him with both barrels loaded.

"Wait a minute," I said. "Looking around here, I do not see anyone who pays my bills or takes care of me when I am sick. In addition, I do not see anyone here I have to answer to and that includes you, big boy!" There was no answer on his end so I thought maybe now would be the perfect time to end this. I asked him not to call me anymore, that I was tired of him and that he bored me. I slammed down the telephone, hoping he would take the hint this time. The phone rang three more times with Rodeoman getting the answering machine. He knew I was here and literally begged into the answering machine for me to pick up the phone, but I was through with that one. I would not be told what to do and would not answer to anyone at this point in my life.

Surfing around the Net, reading the news, I received a PM from Dodger101. "Hello, my Love" his message read. Smiling at his sweet words, I replied, "Hi, honey," sending a big hug to him and then clicking voice chat on. Oh how I loved voice chat with Dodger's sexy accent. His smooth style and his manly Irish voice had me melting into my chair. With his strong Irish brogue, he asked if I wanted to play and I answered that I did. Grabbing my headset and arranging the extra line, I positioned myself while he spoke to me in that sweet Irish voice. He was telling me how my sexy, slow southern voice turned him on. Oh tell me lies and make me like it, I thought with a giggle. His breathing had changed now as he spoke in whispers, which was just what I needed to get the old imagination pumping. He began saying how he was kissing my face and my neck, working his way around and down to my breasts. He was saying all the right things to me but my mind was slipping away to other places and times. My toy was vibrating strongly as I rolled it over my naked body. Messaging it over each breast and around the tips of each nipple, my breathing became rapid. I could hear him talking about fondling each breast carefully, telling me how he loved the

73

way they felt. I spoke back to him in low whispers, even gasping at times. I returned his kisses passionately and was covering him with my hot mouth, licking and sucking his body violently. All the time thinking of another, but with my eyes closed it made no difference—all I needed was a live voice. Soon, I could hear his moaning getting louder as was mine, and as we both began to cum, our voices raged into our microphones. Giggling and laughing as if we had just won the main prize, he told me how special I was. Still, it was nice to hear even if I did not believe a word of it. The main goal of cybersex was not to fall in love but to relieve yourself sexually—at least that is how I looked at it. Dodger was a sweetie and I was sure he liked me as a friend, but he was too much older to think there could be anything between us except friendship and maybe good sex on the Internet. We talked for a while longer about family and life in general before saying our farewells.

I wanted to surf again and check on my online coffee orders and read some more articles before attempting to get some housework done. I had not completely unpacked my suitcase yet. The Internet was my savior from life but also my torment. Just as any addiction, I never seemed to get enough of it. There was too much to see, too much to experience, and the world was at my fingertips. In the real world, I ignored friends and my own children most of the time. My beloved pet would be fed with a quick pat to his head, and even he was suffering from lack of attention. Guilt always seemed to sweep over me but would disappear quickly as another friend sent a PM or web page for my bookmarks. I was like a sponge, soaking it all in and going nowhere with it. In all fairness to myself, I had learned many new and valuable things, but my problem was never getting off the Net long enough to experience them. Jim was an exception and I missed him so much, but I had things I had to do first.

I spent the next four hours cleaning my house and even going outside to play with the dog. As I was about to stop for a break to watch some television, the telephone rang and it was Jim.

"Hi, honey," he said, and a smile surfaced on my face for the first time in days. Oh I was so happy to hear Jim's voice and I knew I had missed him. He asked how I was doing and said he had just come online and had not seen me. I explained to him that I needed a break and my house needed attention, making him laugh. He told me that he'd sent me a card in e-mail and, hearing that, I connected and was online going to my e-mail box. His card was there as he had said and, opening it, I almost cried out. Red roses everywhere, with the words "I love you" flashing across the screen was his e-mail greeting.

"Oh, Jim, thank you so much for the card," I whispered in my excitement. I listened to his voice saying, "Veronica, I love you so much. I wish there was a way I could make you love me back," he said. "I will give you everything your heart could ever want. I'll make you happy, darling," he continued.

"Damn, Jim, you know I would give anything if I could tear my heart out and replace it with another that could love you in the way you deserve to be loved, but right now I can't do it." Feeling very low at that moment for having said that to him, I continued talking. "Jim, I have been honest with you about my feelings and you have known for a very long time what my feelings were about all this. It breaks my heart so much sometimes to know you're hurting." God, he was crying. I just could not say anything else to him, and then he said he would call later and hung up.

What a fool I was walking around the house mumbling to myself, calling myself stupid. I looked at my computer, turned it off, lay down on the couch, and cried. How could I love someone who I knew did not love me? Someone who had stolen my heart and then handed me back the shattered pieces? Somehow and someday it would be okay I knew, but for now, I was going to take it one day at a time. What else could I do? My emotions were going up and down like a merry-go-round. I knew the only reason I did the cybering with the "no face" men was to show my anger for feeling guilty. Sure, I knew it was stupid, but I had to mask myself. I had to cover the pain that was as fresh as yesterday's air.

I had no idea what would happen with Larry/Ed but I knew I had to find out—win, lose, or draw I had to see. The fight would be continuous for me until I could finally drop it once and for all. I had no time frame on when that would be.

I went on a search for new web sites with chat rooms to play in. I was shutting down thinking I was through for the night, when my PM screen jumped up with a great big hello from a stranger called "williestudd." My first reaction was, damn, what a stupid name, but this was the Internet. Everyone's ideas were different about the names they wished to use in chat.

Willie and I chatted for a few minutes. He was an older man looking for a little companionship, someone to talk with, but then again, many men started out our conversations this way and I was in no hurry to push him. Frankly, I could have cared less, but I ended up letting him add me to his friends list before saying good night to go surfing again. There were some sincere men on the Internet, some that honestly were looking for companionship, but I could not be everything to everyone. Sometimes I wished I could help more of the nicer ones but I had to remember that there was only one me and I had only one heart to share.

Deciding to visit back in my old home chat room, I found my old friends there waiting. I am not sure if these ladies cyber or not, but they were nice to me and it was none of my business either way. I enjoyed their hugs and hellos even if some were not authentic. Still, at times, from little bits of conversations, I would say that the biggest part of them did cyber. Like me, they kept it hidden the way it should be. It was no one's business what went on; it was all private and a matter of choice. All of the old room regulars were there and we were having a great time playing music, play acting, and flirting with all the men. I was bombarded with eight PMs flashing, wanting to play with "LadyHawk." I canceled them out and disabled my PMs so that I could play in peace for a change.

All that night I had been watching and waiting for Ed, seeing him bounce in and out of chat, but now all of a sudden he did not have time for me. I knew that this was from my being in such a hurry the last few times he had PMed me, and I smiled because I had expected that from him. So much like Larry, I thought to myself; so hot for conversation and showing attention then when things do not go his way he pouts and acts silly. Well, so be it, I thought, you certainly cannot lose something you do not honestly have anyway.

At this point, I felt sorry for him. Maybe I felt sorry for myself as well because I had hoped things would be different this time. I find myself wishing now that Larry and I had done the cybersex thing when we were an item. I had been his lady and he had placed me too high for that. If I could have stimulated his sexual side, perhaps I could have shown him that I was the only woman he would have ever needed. Could it be that I loved a dream?

The facts were that Larry/Ed was a mere shell of a person, a shallow character with no depth. I was the fool of the two because I was the one who was in pain, not him. He was not a player because even a player has some morals. He was just a common liar with no standards, just as he had always been. For these reasons, I did not believe I wanted to talk to him anymore. What was the sense of it anyway? I could not read his mind but I knew he was not a real man from his actions on the Net. He was not my hero at all. His attitude was always one for himself, not caring who he hurt or how he hurt them; that was very typical of him. I think he planned coming into the 60s chat knowing that when I saw his name, I would be curious. I guess what he did not expect was that I would remember a lot more than he counted on. The biggest thing was his picture—which was a dead giveaway. I did not have one of him now but I would never forget what he looked like. Let's face it, if you love someone, you damn sure remember their face and their voice.

The question that haunted me was why he'd spent so much effort adding me to his friends list, then playing as if he was someone else. Could it be his way of holding onto some past memory? I did not know any longer. I wanted to see where it would end up but I knew I had to protect myself. In the back of my mind, I felt like I should tell him to take a flying leap, but my heart said wait. However, I was not playing the cyber game with him any longer because he was not good at it for one thing. In addition, because I felt that he was such a low creature, it made me feel dirty to deal with him at times. God works in mysterious ways, so I would just continue to play the part a little while longer. I had turned down a wonderful man just to be near this jerk so I needed closure this time. You never know how a story ends until you finish reading the book, I thought to myself.

Checking my e-mail later that night, I found a message from a man I had met in chat a few weeks back; he was from Oregon. I had to smile as I read the letter. It was very polite and very sweet. Acting like a schoolboy, asking for a date, he had me smiling as I sent him a reply. "Sure would love to chat with you sometime when you are in Yahoo chat again," I typed. "Just send me a PM when you are ready because I am always here."

One thing about the Internet is never having a short supply of men who were searching for a special lady in their lives. I would say it was the same for women, too. The key was to remember that people are people, on or off the Net—some good and some bad. As in life, everything was a chance. The Internet just makes it easier to reach out and touch someone.

Checking my other e-mails, I knew there would be one from Jim. He'd sent one every other day as he had done since we first met. I admired him for his relentlessness, courage, and for hanging on to what he believed. I sent my reply in a beautiful card with the words that I missed him more with each passing day. Now that was beginning to sound better each time I said it.

I had barely sent Jim's letter off when I started getting a request to add Thomas to my friends list. Stopping to think, I remembered that this was the man from Oregon and he was fast, too. I giggled as I accepted his request, but I had to surf the Net to check out some new sites and told him that I would catch him later. The one thing not to do is show aggressive behavior. Always keep them guessing and that was the one thing I had never done with Larry/Ed.

Thomas seemed like a very nice man and we talked for about two hours. We just seemed to open ourselves up to each other like old buddies. We talked about music, our likes, dislikes, and we even talked about the Vietnam War. Thomas never once said anything sexy to me and I was glad. I needed friends right now, not a lover. Saying our good nights ended the PM and I decided to look for another online coffee company. If I did not go out to feed the dog, I would have never left the house. My hermit status was building to higher grounds now and that did not make sense to anyone but me. I did not even watch the news anymore. I could not give a flip about the news. I figured out a long time back that it was a rat race out in the real world—and the rats could have it.

That night after I shut down my computer and went to bed, I ended up having nightmares. For some reason, I was being chased by this great big monster computer that kept telling me in a very mechanical voice that I belonged to the Internet now. Maybe I should not have watched the movie *The Matrix*, but I woke up in bed sweating many times that night. Finally, I sat up smoking cigarettes and thinking about things. I was always trying to rationalize and analyze what was going on. Could it have been my inner-self telling me that I needed to get a real life and live again? It was too bad the dream did not tell me how I was supposed to do that. Actually, I saw nothing wrong with my lifestyle. My children were grown and my boys always dropped in a few times every week. My daughter was living in Vegas and was sending me pictures of her belly as it swelled with her first child. She was thousands of miles away but I could voice chat with her for free and could even send pictures. The Internet has many good advantages—perhaps I was using it wrong. Laughing, I knew I was using it for my own selfish purposes. Maybe there was no real love to be found on the Net. Who knows. I had always been one to try anything one time and if I liked it, I tried it again. The sex part often got me a wee bit down, but I did not have any regrets about it because I had learned some valuable lessons, too.

My mind just would not shut off no matter what I did. I thought I was too old, too in control to have nightmares, but I discovered that was simply not true. I took my pillow to the couch and, clutching it like a child, finally fell asleep for a few hours of peaceful rest. Maybe I needed the comfort of another soul beside me or maybe I needed stronger sleeping pills, I thought, as my eyes finally closed.

Turning on the coffeepot the next morning and rebooting my computer, I sat there trying to organize my notes. I was forever writing things down during my online hours. A normal night left my desk covered with papers, food, and filled ashtrays. I was thinking about my conversations with Thomas, Jim in all his sweetness, and my sexy Dodger. I looked at my mess, shook my head, and headed for the shower. I was already dreading the day because I knew I was going to end all conversations with Larry/Ed. I had already made up my mind to this but I knew it would not be easy either. I'd played the game and far as I was concerned, it was another loss. Larry/Ed had already started showing signs of the past, meaning he had not changed in any way. I knew I had to cut him loose or he would end up tearing my heart out again.

Letting the water hit my body hard, it was my time to rethink things. My thoughts focused on Ed as always. "God help me be strong enough to forget," I said out loud as I hit my head on the shower wall. "Let me be free please," I whispered. As my tears flowed, I wondered if I would ever be able to let Larry/Ed go. Something deep inside of told me that I could never love another man like I had loved Larry and I should just be happy that I had known real love in my life. Yet, that also was easier said than done. Physically, yes—mentally, maybe never, but somehow I had to try. It was driving me nuts.

Turning so the hot water was hitting my face full force, I hoped that it would drown out his voice and his laughter in my head. Stop it! I screamed at myself inside, stop reliving all that hurt again. Taking a deep breath, getting out of the tub, I grabbed my robe and headed for my coffee. I had left the computer on standby all night so it was ready and I was trying to be.

My mind would not stop, I could not help but think of all the lies he told me. I kept thinking how confused he must be too. It has to be hard to tell so many lies and not get confused at times. He said to me in one conversation that I had a bad attitude when actually he was the one with a

bad attitude. How much hurt and disappointment could I
go through again? I asked myself. The hell I put myself
into was of my own doing and I just didn't know where to
go with it. With every word we exchanged and every
message we typed I was in pain. Maybe it was time to end
this charade and maybe today was the day, I thought, but
deep inside I was lying to myself. For my own well-being I
had started talking to him with hope of a renewal and then
it turned to revenge, now all I wanted was to be free from
the lies, both his and mine. I was tired of it all.

Surfing the web pages I knew that if I stayed online
long enough he would show up. Whatever he was looking
for, it was not me as much as it grieved me to even think
that way. I was having such a hard time facing this and I
didn't understand why. It is hard to give up the only man
you have really loved, the bitter sweetness of dreams that
have failed. I was a woman who needed complete love. I
needed to feel secure in that love anything short of that
would never make me happy. The idea of never turning on
my computer again went through my mind, too. Again, I
just didn't have the strength for that now. I hoped I could
be the old me again, maybe even becoming the happy
lady I once had been, stopping the pretense.

As I knew it would happen I watched as his name
came online and, waiting a few minutes, I started typing a
message to him. I couldn't wait, I had to do this fast or I
might not do it at all. It was now or never for me.

"Ed," I typed on my PM screen, "I have had enough of
all this fun and games. Let's just say we had fun and move
on." Looking at that PM, I hated to push send but I did and
then I closed the screen out. My next move was to delete
him from all my lists and put him on ignore. Leaning back
in my chair, I knew I would miss seeing his face light up,
but at least now, I would not be one of his victims. Being a
victim meant that I gave him control over my life and this
man deserved no control over me. So Ed, I thought, or

whoever you are now, God Bless you, you crazy son of a bitch, and God help you and the ladies that you sucker into your web. I am just glad I can be free again. To top it off, I felt very relieved. I would carry the sweet memories of the past in my heart and do my best to forget the bad.

Turning off the computer, I got dressed and headed out to shop and to get some fresh air. Later, I would talk to my friends and I knew they would be there for me.

Katie was not my only friend, I had another great friend, Gwen. I was always thanking God for giving me two such wonderful people as friends. Gwen was just a few years older than me but seemed so wise. I felt like she was an older sister and I valued her friendship enormously. She was also one of my computer teachers. If I had a problem with a program, she was there to help and would not stop until she knew the solution and until the program ran well. Sometimes I felt that if it were not for them, Katie and Gwen, I would have crawled away to die. They were there for me as true friends were and I loved them both. No matter how crazy I got, they were always there for me.

Chapter 9

Of course, there were many good people I knew from my home chat room. The Kat always made me laugh, a cheerful lady who thought she was a cat and scampered across the screen purring and bouncing around. She was just plain fun to watch. My friend, Pond, was so smart when it came to computers. I considered him my friend and another one of my teachers. Plus he was such a sexy devil, but because of our friendship we had never cybered, we only played with words.

There were many good people in my homeroom chat, I thought, as I listened to the music roaring in my ears. Letting go of my thoughts, I started to surf again. Checking my e-mail, there were at least five cards from Thomas, and two from Jim. Jim would be happy to hear I was finally setting the past down, but I still needed some time to pull myself together. I could not be sure I ever wanted to open my heart to another commitment and until then, he would have to wait. I was still a player as far as I was concerned and I wanted to play.

That thought had no more than passed when my friend Dodger appeared and we started chatting. When I heard his Irish accent, my blood started to rush. He had told me that my voice did the same to him. We talked for about ten minutes when he whispered how much he had missed me and I laughed knowing what he'd missed and to tell the truth, I had missed him, too. Going into voice chat, the first thing I heard Dodger say was, "Oh me love, I wish we were closer so I could hold you and make love to you the way a man should."

"Oh yes, that would be wonderful," I said as I grabbed my toy and adjusted my headset. Laying down on my couch, I listened to him talking about touching my body as I touched it for him. He talked about how he was kissing my face and my throat. There was never much to say on my part, he was the one who led and took control and I would close my eyes and listen. I could hear his breathing getting faster as he spoke of my breasts being caressed and kissed. His hands were now exploring my flesh and I was whispering back to him about touching and rubbing his cock. That was all it took for him before he was opening my legs, kissing around my stomach. My darn toy needed new batteries, I thought as the vibrating slowed. But I loved the way he talked about leaning over me and pushing himself into me. His breathing becoming faster and faster as mine did. With my eyes closed, I moaned into the microphone and knew he would simply go mad. It is what we call here in the US, a quickie, but that was all I needed and if it worked for us then it did not matter. Soon, his breathing got very hard and raspy. He started talking again and asked me if I was ready and oh yes, I was ready. For some reason, it felt more intense this time and I almost screamed, knowing he could hear me. I was lost in my own moment, my own feelings, as my body gave away to the sexual release. Sometimes I wondered if my heart could stand the pressure of these acts but I always seemed to come back to earth and I could hear him reaching his point, too. I whispered into the mic softly, "Cum for me baby, do it now!" With that, he released, too, and with a smile, I knew that this had been a success once again.

Still laying very still, trying to get my breathing to slow down, I heard him say, "Jesus, Joseph, and Mary woman, you are just too much," and I giggled at him. Actually, I had done very little in this game, but if it was important to him to think I was wonderful then why should I rain on his parade? Laughing was always part of our finishing. I laughed at him as he tried to get himself under control, too.

"Good Lord, me lady, you are just so grand. I want to marry you and do this every day in person."

"Oh yes, my friend, but I am afraid at our ages we would surely kill each other," I said and we both laughed. We chatted another ten minutes before I decided it was time to stop and get on with the day.

Deciding that I wanted to take a drive, I jumped in the car and went out to the lake that was nearby. I wanted some fresh air and there was something very calming about sitting by the water. Looking across the tiny ripples, I recalled back to a few years in the past when I lived in Kentucky with my husband. We were both involved in a local wildlife center and I loved it. The animals were great and I even helped raise a few of the smaller ones, playing mama to many of the ones that had been abandoned or found lost. It was a great feeling being a part of Mother Nature's plan. I sure missed it and my old friends there. I almost hated that I left but sometimes with a divorce, many things change. Making a mental note to call my old friend Ken when I got back home, I took one last glimpse of the water and headed back.

I had many plans today for my computer and I. My neighbor wanted me to get her set up online so I bought some food and finally made it back. Besides talking to the men, I did lots of surfing—keeping up with earth happenings at Discovery.com and learning the Cherokee language from another site. My time was well planned and I wanted to keep busy. Stopping before I came back into the house, I pulled a ball out of a sack and threw it to my friend Rebel. He jumped for joy and for the ball. Throwing the ball into the yard, I enjoyed watching him run after it. Playing with him for about fifteen minutes was all either of us could handle. He refused to go fetch the ball and I was ready to sit down feeling satisfied that I had spent some quality time with him. It was now computer time again.

It was difficult for me, not looking for Ed while I surfed the Net, but I was holding my ground. The man was obviously only interested in his own feelings, so screw him,

I thought. Talking to myself seemed to help me cope. Going into my chat room, I was determined to keep my mind focused.

There was the gang all hugging and saying hello; someone was playing some great songs and I just sat back and laughed at all the silly goings on. I must have been there about ten minutes when I received a PM from Thomas. That one PM lasted three hours and when we finally said good-bye, I felt great. So far, he had only gotten romantic with me and I loved it. This went on for about two weeks and each PM seemed to increase my curiosity about the man. He would get the conversation heated with little comments about how nice I was and how he would love to hold me and I would grow silent. When that happened, he would apologize and we would start talking again. Although I had cybered many times before, something told me no with this man, and I was not sure what it was but I was going with my gut feelings.

All my attention was being focused on Thomas. Looking back, I can see now that it was my way of running away from Ed, possibly from all the men I'd played with. I did have a good time with most of them, flirting and carrying on. Maybe I was looking for the ultimate hero, but I think I knew there was not one out there for me. Still, life goes on and I wanted to live each day and give it my all, even if was just on the Internet.

I had barely opened my e-mail box when Thomas sent me a PM. That man must have been waiting on me because that was just too fast. Laughing, I sent a hug asking what he was doing and he said he would like to call me and asked for my number. Since we had been talking for a couple of weeks, I decided that it was okay to give him the number and within thirty seconds, the phone rang.

We must have talked for an hour and I finally figured out what made me uneasy about him. At times, he sounded almost spastic, too urgent to have someone in his life. I did enjoy his attention though, so we talked and talked. It was not long before he wanted to plan a meeting and that made me nervous. I did not say no, only that I would think about it and he seemed happy with that.

While Thomas and I were talking, Rodeoman sent a PM asking if I was busy. No, I was not busy since I could type and talk on the phone at the same time. Rodeoman got right to the point by saying how much he had missed me but that he had been out of town on a job and now he wanted to play. I did not have the time right then, so I thought of a good lie because he would have never accepted a simple no. I told him I was just leaving my house to visit my dad at the hospital and maybe we could get together later. I could tell from the next PM that he was not happy but I didn't care. It seemed like the more I talked to "Rodeo," the more agitated he became with me. Damn, I did not have to put up with that crap so I said good-bye once more and closed out my PM screen, returning my attention to Thomas. Thomas had been talking a mile a minute the entire time I'd been typing.

"Wait! Where did you say you wanted us to meet?" I asked Thomas.

"Can you meet me in Chicago if I send you a ticket?" he asked me again.

Wow, Chicago, I thought. It had been many years since I had been there. Now that was an offer I could not refuse and I told him to send a round trip ticket so I would be assured of a flight home, that I would be happy to meet him.

"Coolbeans!" I said out loud. I was going home again. I had been raised in Chicago and the idea of seeing it after twenty something years made my day. Thomas agreed with the round trip ticket and before I knew what was happening, he was planning our meeting in two weeks. I was happy and bouncing from the thought of it. Thomas and I talked about ten more minutes, then said our good-byes for awhile.

I was in a very good mood thinking about my trip and decided that I wanted to cook. Being alone all the time, it was easy to run out and grab something but today if I could remember how, I would fix my favorite dish—liver and onions. For the rest of the afternoon, I watched television and practiced speaking my Cherokee vocabulary while my great smelling dinner simmered. Every so often I would jump online and check my e-mail. There was always a card from Jim and now Thomas was sending one every other day, too. I made a note to be sure and call Ken later on. For some reason, I was happy but I was not going to try to figure it out now. I would just ride with the flow.

That evening I had the best time online that I'd had in ages. Even when I think of it now I laugh out loud. Having finished cooking my dinner, I settled in front of my monitor surfing as I tried to eat. From out of nowhere came a sexy message from Bob, my phone sex man, but I did not want to get into the phone thing so I said let's just play here in PM. Bob was all for it so he asked what I had on. Looking down at my sweat pants and floppy shirt, I typed back that I was wearing my long, black gown after finishing my bubble bath and the scent of wildflowers surrounded me. Taking a bite of my liver, I grinned evilly. Running to the kitchen for more tea, I waited for a response. The PM screen came up again and he was talking about holding me and kissing me all over. I needed more salt so I made another dash to the kitchen. Sitting back down just in time to reply, I typed back that I was returning his kisses and running my nails along his neck and chest. Hitting the send button, I thought I should cook more often, the liver was great! I must have typed in every sexy phrase I could think of while sitting there eating my liver and onions, laughing my ass off and spewing tea all over my monitor the whole time. Finally on Bob's last, "Oh baby" I could not stand it any longer, I threw my headphones off and literally got on the floor and laughed so hard I nearly did not make it to the bathroom. When I finally made it to back to the monitor, Bob's last PM was there saying his wife had come he so he was out of there. Crying from all the laughing, I thought, God bless the Internet. Where else could I have so much fun while having a great meal? To this day if I disappear in chat for any length of time, my friend Kate will PM me asking if I am having liver and onions again. With that we would laugh our butts off. There is never a dull minute it seems in our chat rooms.

After I had settled down and cleaned all my dishes, I decided to call Ken from my computer. Dialing the number, I could almost smell the freshness of a rainy day in Kentucky. If my fate had not gotten twisted there, I would not have left. Kentucky has plenty of fishing, lots of water, and a lot of wildlife if you lived in the rural areas. I was sitting on my front porch one day when wild turkeys came out of the woods. It was grand and then there was Ken. Ken was one of the good guys and a professional naturalist. He was one of the best Herpetologist in the country as far as I am concerned, and one of the best friends. He was a true gentleman and always made me mad because he knew everything. I knew him well enough to ask him to explain to me in idiot terms about things we were discussing and sometimes he got carried away. Giving him that LadyHawk go to hell look, he would always grin, waiting for my comeback.

We talked about everything going on at the wildlife center and all the new projects. I had set up his computer system, but he still did not get online much. It was so good to hear his voice and then he said, "If you want to come back there is a trailer nearby for rent or you can move in with me." It was just a friend offering a friend a place without any hassles. Laughing, I thought he was like a twin brother reading my mind. That was a thought I had had for some time now, but what scared me was the lousy Internet Servers in the area. Being without online access would never work. Saying our good nights, I felt so much better just talking with him. It felt good to talk with a man you could be yourself with and not worry about putting on any airs.

Feeling better, I started surfing the Net when I received a PM from Thomas telling me he had information about our Chicago trip. Naturally, that got my full attention. I had come to a realization that I would do anything and go anyplace if it would help me forget Ed. I asked Thomas to call me on the phone just saying I needed to hear his voice. My phone rang right away and he started telling me he had my tickets and they would be arriving in three days. I would leave for Chicago on Friday and would meet him at O'Hare airport and from there we would go to our motel. We were going to be there for three days yet he only mentioned us having separate rooms the first night. I guess the other two nights were playtime for us. He had spent a bundle on this trip and I wondered where the extra money for all of this was coming from. Instead of me flying to Oregon to meet him, we were both flying to Chicago and would meet there. Something seemed out of place with his plan, but I could not put my finger on it. I thought it was me and went on about my business.

Thomas must have called ten times a day after that and he was making me nervous but I was packing and getting everything ready. We had talked on our Intel cameras so we knew what to expect as far as looks. My kids did not like the idea of me going to Chicago, but they hadn't liked me going to California either.

Friday rolled around and my son took me to the airport, fussing at me all the way. We got out of the car and he carried my bags into the airport, then turned to me and said, "If you need me, call." We hugged good-bye and then I was flying those friendly skies.

The flight did not last long, but long enough for me to have two drinks on the plane and after that, I did not much care what happened. If you do not drink, two drinks can hit you very hard, especially on an airplane, so I was departing my flight with a giant smile on my face.

Even with glasses my vision was a bit blurred and there were many people all around, but I was in Chicago and taking a deep breath, it smelled as I remembered.

Hearing my name, I turned to look and there was Thomas waving at me from across the crowd. Smiling, I sat my bags down at my feet thinking there was no way I was going to fight this crowd so I let him make his way to me. Within a few seconds, he was throwing his arms around me holding and kissing me. All I could think of was that he needed a breath mint. Picking up my bags, he talked a mile a minute as we made our way through the crowd. When we were finally outside, he hailed a taxi and we were off to our motel. Thomas was asking me what I wanted to see while we were in the city. All I could think of was State Street. I remembered State Street with all the lights. I did not think we had time for the museums. I was happy to be here, and it was almost like being home again. I was in heaven—or maybe just drunk.

Thomas turned out to be a very handsome man and although I had seen pictures of him, I thought he looked much better. Maybe this trip would be worth my while, I thought, chuckling to myself.

In the taxi, I watched as we drove along the Loop. I could almost see myself riding my bike along the walk that I recalled from my childhood. The city had changed some over the last twenty years or so and certain street names I remembered while others I could not. We were in the downtown area when our taxi pulled to a stop in front of our motel. Being in awe and being tipsy had a lot in common, I decided as we checked in and then went to our rooms. Just as Thomas had promised, I had a room to myself. He was being such a fine gentleman. He helped me settle in my room and then took me in his arms, holding me for the longest time. He let go saying he had to go take care of business and that I should relax and he would come back in a short time to take me out to eat. Business? What business, I wondered, but I was not going to worry about it. What ever his business was, it was his problem. I noticed that he was still carrying a very large briefcase around. I had noticed it at the airport but had not given it any thought then.

After he left, I lay back on my bed and looked out of the half-opened doors leading to the balcony listening to the sounds of the city. I must have dozed off for suddenly, it was dark and Thomas was lying beside me softly shaking me.

"Are you ready for dinner?" he asked. I started changing my clothes and putting on fresh make up. All the time Thomas sat on the edge of the bed talking as if he could not stop. Bouncing in and out of the bathroom, it suddenly hit me. This man had talked about all this wonderful lovemaking and yet all he had attempted to do was to hold me and kiss my neck one time. Shaking the weird feeling, I listened as he called a taxi and then we were heading to State Street to eat.

Oh my God how I loved seeing this street again, not much had changed either. I did notice that the number of prostitutes and homeless people seemed to have increased with the passing years and somehow, the city looked dirtier than I remembered from my childhood years. Zoning back in to what Thomas was saying about food, I thought it did not matter where I was going to eat, I was just very hungry. My mouth watered for a Chicago hot dog but Thomas told the driver to go to Papa's Little Italian House just off State St. Okay, I thought, Italian was great, too. Thomas was still carrying his briefcase and I thought that quite odd since we were going to dinner.

Maybe it was my nature to be curious about such things but I did not want to ask him why and take a chance on spoiling the evening. Another thing that bothered me was that he seemed to know too much about this city. He said he had never been to Chicago before and had only surfed the Net for information about motels and restaurants. I guess that could have been right, I thought, as we arrived at Papa's. The building was dismal looking outside but the inside was very nice. There were small tables with candles, the lights were low, and the smells coming from the kitchen were marvelous.

We were seated at the back table in the room, which seemed to be what Thomas wanted. When I saw the male host wink at Thomas, I grinned, thinking he was perhaps gay. I then began to wonder about Thomas. When we were ordering, I was shocked to hear the waiter and Thomas speaking fluent Italian to each other. When the waiter left, I told Thomas that I had not realized that he could speak Italian and he just smiled at me, pulling himself closer to kiss me on the cheek sweetly.

The dinner was wonderful and the wine was going straight to my head. We were laughing and talking and about to have dessert when the waiter came over and whispered something to Thomas. Thomas said for me to enjoy my dessert, that he was going to talk to an old friend that had stopped in the kitchen area. He picked up the briefcase and walked through the kitchen doors. Now I may have been born at night, but it was sure not *last* night and something did not seem kosher here. Things were not adding up at all and my stomach was becoming all upset. When the waiter came with the dessert, I asked where the restroom was and he pointed toward the back. I was about to open the bathroom door when I heard Thomas's voice from down the hall. I tiptoed a few steps closer to see if I could hear what was going on. The only thing I heard was Thomas saying something about $50,000 dollars, then all hell broke loose.

Doors were being kicked open and men with guns were rushing everywhere screaming for everyone to get down. Customers were screaming and I was pushed onto the bathroom floor and ordered to stay put. In my confusion I thought he had a gun!! Holy Mother of God what was happening, I wondered as I started to pick myself up off the floor. Some man dressed in black burst into the restroom, looked around, asked me if I was alone, then pushed the stall doors open looking with his gun ready. I wanted to tell him I had been going to the bathroom alone for years now, but I knew when to keep my smart mouth closed. All of a sudden, I saw Thomas being lead away in handcuffs and I called his name in disbelief. He looked at me and said, "Well, those are the breaks, darling." Another man was asking if I was with that man and stupid me said, "Yes, I am." Needless to say, I was then pushed against the wall and placed in handcuffs, too. There were four men asking me questions as they pushed me out of the restaurant door and into a squad car. I thought I was going to go into shock a few minutes later as I was being shuffled into the 59th Precinct. It took about three hours for me to find out what the heck was happening. Two plain-clothes men took me into a room and asked me a million questions about Thomas and I told them all I knew. They took the handcuffs off, then told me to sit still while they ran a check on me. Well that would be interesting, I thought, but even my own humor was fading as I tried to light my last cigarette.

I had been sitting in that room for two hours when another officer finally came in. He said they were going to hold me overnight for more questions and that I would have to go to a cell. I panicked. "Don't I get a phone call?" I asked. They always do in the movies! He took me to the pay phone but I had no money for a call. My money was hidden away at the motel in my socks. Trembling, I dialed the number to the only one I knew who could help me—Jim.

Linda answered the phone, I told her who it was, and asked her to tell Jim that I was in serious trouble. Looking down at the card the first officer had given me, I told her the name and the number on it and told her to get the information to Jim fast. Then I was pulled away from the phone and taken to an empty cell by another officer.

My feet were killing me from all the pacing around that stinking cell. Nearly another hour passed before two more detectives came to talk with me. Opening the cell door, they walked in looking mad at me. I wish I had donuts to give them, I thought. They started telling me that I was in serious trouble and that I needed to tell them everything I knew, so I started talking. I told them how I'd met Thomas on the Internet and that this was our first meeting. I told them that it had been his idea to meet here in Chicago. The short one asked me if I could prove that and I said sure, if you have a computer I can use. I was taken to a regular PC where I went into my online e-mail box and showed them my letters from Thomas. The short man took over and went through all my e-mails. I did not want to stay here in jail, all I wanted to do was go home. I promised God I would never stray from home again if he would just get me back to Tennessee. The short guy began to print my e-mails and I was taken back to the holding cell.

Sitting on the bunk, which I now called the rock, I waited. Another hour passed and I just could not make myself lay down on the filthy bunk. Soon, two more detectives came to take me for more questions but at least these two were nicer than the last ones and gave me cigarettes. This time, I was escorted into a much nicer room. It was five o'clock in the morning and I was beat.

The detectives went over my story again and this time I asked just what was going on with Thomas. They explained that Thomas had been associated with the mob for years and had been watched for a long time. He was a drug supplier and after searching the briefcase he'd carried so securely, they had uncovered thirty-five pounds of uncut heroin, which I was told was supposed to sell for $90,000 dollars with the street value being over a million dollars. All I remember after that was that I started to get sick to my stomach and my head started spinning. I must have fainted because I woke up sometimes later in another office and Jim was there holding my hand. When I saw his face, I grabbed for him and started crying. He held me tight whispering to me that it was okay; he had everything straightened out and we were leaving. He helped me up then held me until I steadied myself. I had to sign a statement before I left the police station with a warning from the top detective that I should be more careful about the people I meet from the Internet. Looking him directly in his eyes I said, "No shit!" Jim laughed and so did the detective. I leaned on Jim, exhausted and scared. Jim did not say much to me about it after that. He held on to me asking me if I was going to be okay and I could not stop thanking him enough for helping me.

When Linda gave him my message, he'd gone directly to the airport and caught a flight to Dallas and then one to Chicago. Now this was a true hero.

When we got back to the motel, I took a shower, collapsed, and went to sleep. I woke later to the smell of fresh coffee and Jim leaning over me stroking my forehead. Smiling up at him, I put my arms out for a hug and Jim slipped into them.

"Come on," he said, "drink your coffee." There was fresh Danish, too. We talked over our coffee for hours and he laughed at me for getting myself into such a mess. But that was my life, one mess after another it seemed.

Jim had a late fight the next day and so did I, so we made plans to visit the Museum of Natural History early the next day and then head to the airport. That night, I lay in his arms and snuggled as we slept, but we never made love and I never thought anymore about it. The next morning, we had our coffee again and spent a wonderful day playing in the museum. We ate lunch and talked like we had never talked before and then it was time to leave for O'Hare. My flight was scheduled to leave an hour before his. We hugged some more and said our good-byes with a promise to talk soon.

Heading down the ramp to my plane, I stopped and turned around to see Jim standing there watching me. Smiling at him and feeling suddenly very sad, I waved and boarded the plane feeling like I was leaving my best friend behind.

On the plane, I did some soul searching about myself and came up with conclusions. One thing I knew for sure was that no matter where I went or what happened to me, I did have a few good friends to count on. Getting off the plane in Nashville, my eyes searched the crowd for my youngest son. When I saw him, I thought about how much he seemed to have grown, but that was the mother in me coming out. Waving at him, I rushed over and hugged him hard.

"Gosh, Mom," he said, "I just saw you six days ago."

Laughing, I told him that I always missed him when I was away. He looked at me strange but I was not going to tell him anything about my adventure. I knew better. If he knew, he would tell his older brother and sister and they would have grounded me for sure. I kept silent and we talked all the way home about his new job and girl.

It was so good to get back to my little apartment again and my friend, Rebel, but most of all, it was good to be back to my computer. Turning it on to warm up, I put my clothes away and made some coffee. I was ready for a day of online adventures. Running through my e-mail boxes, I discarded all the junk mail and read all my mail from friends. Several online messages from Dodger, Rodeoman, and Bob were waiting for my attention, but none that I cared to take seriously at the moment. I went into my home chat room and spoke to a few people I knew, but there were many there that I did not know. New people were always coming in and the room had changed so much since I'd first arrived there. It was not the same place anymore, where we all knew one another and were all friends. As I sat there and watched the people talking in the chat room, I felt very sorry for some of the ladies. Some seemed to beg for attention whether they were married or not. Some were very bold while others were shy.

As in any group of people, there were some I did not care for. I had been in the rooms a long time and I knew some of these ladies like the back of my hand. A few women I would not trust as far as I could see, they were the back stabbing bitches out for only one thing, themselves. Some you could tell were just straight out whores ready to jump anything with a dick that would have them. These women stayed in their own group it seemed and pretty much hung together. It was comical to see them hug and say hello to everyone when you knew they actually hated each other. It might have been jealousy they had for each other, but it was hard to tell. If a new man came into the room, they all suddenly became like bitch dogs in heat. Yet there were some you could not help caring about because they were great ladies.

It was hilarious to watch the room at times and I was well known and was sure some hated me, too, but I did not care. I was not in the chat rooms for them, I was in the chat rooms to play and have fun. As much as I tried to be friends with all, it was impossible. I had been invited to many of the chat meets where many would meet together and most for the first time, but I never felt comfortable being around the large groups. Besides, the few I really disliked always seemed to be going and I knew with my temper and my honest nature that I would be in trouble. I would end up telling them what I thought and then possibly have a physical confrontation. There were a couple of them that I would have loved to slap silly anyway. At forty-eight, I was not afraid to stand up and knock someone silly that got in my face. No it was not lady-like to think like that, but I was not a lady all the time either. Besides, some of the so-called ladies in chat would get in Yahoo voice and use such nasty language it was hard to call them ladies at all. Compared to them I was ready for Sainthood. In fact, I really doubted they were. The word "slob" comes to my mind. I was always what I had to be at the time. I could be a fighter if need be but I would rather be a friend. Then again, you cannot be friends with everyone.

I still had not heard anymore from Ed, but had a sneaking suspension that he was already back in our home chat room under another alias. Not only was he hiding from me but I knew he was also hiding from a couple of others in our room. As much as I hated to admit it, I knew he was a gutless wonder. Ed, you see, was trying to find someone close to him so he would not have to travel very far. He went after any woman in the northeastern part of the country. I had watched him do this many times and then laughed at his feeble attempts to flirt. Why in the hell did this man have such a hold on me? It was a question I asked myself over and over again. All I could figure out was that I was in love with a memory, and that is harder to get over than a real live romance. In my heart, I knew that he would never find a woman that held as much passion in her heart as I did, but I could not make him see that. I was confused as to whether or not I even wanted to try anymore.

I think all my cybersex and traveling was not to find love but to help bury the memories I carried around. Sometimes I did not think anyone knew the real me and maybe that was just as well.

Surfing into my bookmarked pages, I was studying my Cherokee language when I received a PM from my Dodger.

"Where you been?" he asked with that strong Irish voice. I knew I was not going to tell him the truth about Chicago so I told him that my computer had been down a few days. He told to me how much he had missed me and how much he had been thinking of me. While he was talking, I was arranging myself for what I knew was about to happen.

With my slow southern style and my nasty New York mouth, I began to talk about how I had missed him, too. Lying back on my big couch, I listened to his sweet talk. Oh he was now in love with me, every waking minute he thought about what it would be like to really feel my body; to kiss my face and my neck. I could hear his breathing getting faster and I opened myself up. Whenever I cybered, I had only one person in my mind that I was making love to—a beautiful memory. Dodger's voice was almost magical that day as we touched over the air space between us. My body was responding to his voice as my breasts heaved and my body shuddered as my own hands roamed and touched. My voice was strained but I delivered the sexual stimulation that he longed to hear. We played, touched, kissed and sucked every inch of our bodies until finally, I could hear his voice getting raspy and he could hear me moan my own pleasure. There something about a man hearing you moan that really sets them on go. Then all of a sudden I heard Dodger say, "I'm about to cum for you, my love," and I brought myself over the edge with him. Hearing his breathing and laughter, I knew our actions had been successful.

As we began to relax and talk for a minute, I suddenly felt very empty. Dodger kept saying that he loved me and it just did not mean a thing to me. Somehow, I did believe it, that it was not just a sexual thing for him any longer. He continued to talk about our meeting and all the things he wanted to share with me and I played his fantasy along with him. Maybe in his mind he did really love me, but unless we could meet in person, there would never be a snowballs chance in hell of ever finding out. Both of us being very limited on money meant that that meeting would never happen. Wanting to get away from the heart-wrenching conversation, I said I needed to go and hoped we would be able to talk again soon. Moving everything back into place, I was surfing again. I needed to cancel my online coffee orders.

There was no one in my home chat room I cared to chat with so I went on a rampage through some other rooms just looking for trouble—it always found me. First, I made out with a sixty-six-year-old in the "older men looking for younger women" room, then I headed to the "pick up" rooms. I was in disguise under another alias other than my own username. I could not get my mind satisfied that day although my body needed nothing. Just playing with the men seemed to pass the hours until my girlfriends came online again, then I would join them and we would talk like normal women do. We talked about the other ladies online and sometimes our problems at home—normal things and then sometimes not so normal. For some reason, I needed to write down my thoughts. Putting them down in print always seemed to help me cope; I was not sure why.

I was miserable: miserable with him, and miserable without him. What does a woman do, I asked myself again. Then I started writing in my daily journal.

What can I do, what can I say? Is there no tomorrow for these empty arms? Do you not understand the passion I have waiting here? My heart cries but it stands firm to what it feels. No matter what I do or where I go, I cannot rest. My mind never stops telling my heart that you are the one I love. Will there never be another that will ease this pain? I could never allow it. I wander aimlessly in the memories of the past and there is no longer a beam from the lighthouse guiding the lost into the safe harbor of my love. Chances are, I will always be a stranger to you, only holding you in my dreams. I see you smiling at me, singing your little song. I dream of a future praying you will be at my side, longing for the night that you hold me to once again feel the peace that I felt with you.

I closed my journal and sat there crying so hard that I could not breathe. Stepping away from my computer to get some air, my heart started pounding and pains ran down my arms. I was dizzy and had pressure pushing on my chest. Knowing this was not normal, I tried to regulate my breathing, but it was not working. The pain was getting more intense and I was getting very scared. Grabbing my coat, I drove to the hospital, which was only down the block. I forced myself to walk into the emergency room and that is as far as I could go. I slumped down on the floor trying to breathe when two nurses saw me and rushed over, knowing immediately what was happening.

I was rushed into the trauma unit and I must have passed out for a few minutes. When I opened my eyes, my heart doctor was sitting beside my bed writing on my chart. Looking at me he smiled and said, "That was a close call but you're going to be fine."

"What happened?" I asked, though I knew.

"It appears you had a very light heart attack because if it had been more severe, you would not have driven here or be talking to me now. What were you doing when this happened," he asked.

I told him that I had been upset when I noticed the pain. He had been my doctor for a long time and proceeded to tell me as he had done so often that my heart was not able to handle much more. I knew this from my past medical conditions, but it had never actually been this bad, either. He prescribed nitro tablets both morning and night along with my other heart medicines.

He told me that my heart was unable to work on its own when put in extremely stressful situations. With that said, he added, "Your heart muscle is turning into jelly, Veronica and you have known this for a long time now. I want to keep you overnight and repeat some tests again."

"No," I almost shouted, "just give me my prescriptions and I will see you on my regular visit."

Shaking his head, he had asked if I had ever told my family and I shook my head no. "You are a hard-headed woman," he said, "but it's your life. All I can do is what you want," he fussed.

Smiling at him I said, "You know we cannot live forever."

"Well, at least let us monitor you for another hour longer and then you can go home if you feel well enough," he said.

Agreeing to do so, I lay there and thought about things as the machines echoed each beat of my heart.

Finally I was up and out of there and on my way back home, but I did rest that day, lying on the couch watching television and talking on the phone. I had never told anyone about my heart condition, not even my children. My friends and the kids knew I had heart problems yet as far as they knew, it was only high blood pressure. I had told people that I was going to lose weight so by the time I was fifty, I would be a fox. According to the doctor, that would be about all the time I would have. Could this be the reason I secretly did things so intensely? Maybe. I knew a long time ago that I had to live my life to the fullest I could possibly live it. Each day was very precious to me and I wanted to make every one of them count. If I was going to make it to fifty, changes needed to be made within. Sometimes, I wondered if it was worth the effort.

Jim called that evening and we laughed and talked for almost two hours about nothing and everything. I never opened my mouth to him about my heart. There was no sense in him worrying and I knew he would. As we talked, I puttered around the house cleaning here and there. Frankly, I had loved my house spotless but now I did not care as long as the health department did not condemn the place.

After we finally quit talking, I had four more calls to make and it was beginning to get on my nerves. I did not turn the computer on but I knew I would later that night. I mean, I was part of the "Night Shift" crew, so I would be there. I did, however, make one more call and that was to Ken up in Kentucky to let him know that Rebel and I would be up for a few days and I wanted to do some fishing. He laughed at me 'cause it was really too cold to fish, but he knew me and knew that if I was talking about fishing now, then that was a signal that I wanted to be alone to do some thinking.

Cranking up the computer, I went online and into my e-mails, looking at the beautiful cards my guys had sent and thanked each one. I went through all my e-mail address books and deleted all except my closest friends. As soon as I finished that, I headed to my chat room. The room was packed with user friendly names. Some I knew, some I did not, which was always par for the course. After my hellos and hugs were done, I started playing songs, flapping my wings about and dancing all over my keyboard. I had disabled my PM box so I would not be disturbed. I wanted to play and have fun listening to the tunes. One PM kept popping up with the same name but I did not really notice until the sixth time. Looking up to read the PM, my hand started to shake. It said, "Please talk to me, hon."

My first thought was Ed and I hit my ignore button and went on playing in the room. If I ever spoke to him again it would be in person and he would have to travel to Tennessee. Besides, he would only end up causing me more grief, but my heart broke again. My God, I thought, does this pain ever end?

I was getting bored and decided to surf a bit. When I surfed through the chat rooms of Yahoo, there were always loads of PMs. Tonight I was LadyHawk, and I did not need a hero or to chat with anyone. My wings wanted to take me to other universes and plains of existence. I needed escape and release that only I could give myself. Soaring through the heavens at Discovery.com was where I headed. After that, I went to a pagan site on spells and talismans. Glued to my screen, I was reading how to conduct spells for love or any blessing you might want. I thought that was a great ideal, but which one did I want to try? I did not feel stupid for even thinking this. Printing up the pages, I was mesmerized with a love spell I came across. According to the pagan author, he said it was an original piece from a very old Book of Shadows. Well, that certainly sounded like it was another alternative, I thought, giggling to myself. The more I read, the more interesting it became. I realized I could do a spell on just about anything I wanted. All I needed was a special time, a special place, and a few material objects. With these things I could plead to the God's of Old for anything my heart desired—I laughed.

I was drawn to the pagan sites maybe out of boredom, I couldn't be sure. What intrigued me the most were the Native American parts of the mystical beliefs. Page after page of spiritual and religious beliefs from every American tribe there was. I was printing so much of the information I was finding that I went through a few ink cartridges before I felt like I had enough. This seemed very important to me but I did not know why. Taking a break from all the new information I had found, I decided it was time to check my e-mail and let my homeroom know I was still alive. For two days, I'd avoided everyone except Gwen and Katie but I had only told them I was studying my Cherokee language, which was not a total lie. I actually did learn a few words in my new journey.

My e-mail boxes were full and when I opened my Yahoo pager and my ICQ, I was swamped with messages. Deleting the unimportant ones, I noticed a new name and address. It was from the detective in Chicago that had been nice to me. Thomas was now serving twenty-two years for trafficking with the intent to sell and for distributing illegal narcotics. He had actually pleaded guilty and had even signed a sworn statement that I was an unknowing victim. Oh shit, I thought, my name was now listed somewhere in some file cabinet as a victim. Well that made my day for sure. I had to laugh but somehow, it did not seem funny. I replied with a thank you note and I hoped that I would never hear from anyone in Chicago again.

I spent the next few hours answering the messages. My Dodger had missed me, he wrote. Thinking about that wonderful Irish accent of his almost made me go into the chat rooms to look him up. Yet, I still had things to do online. If I were going to visit Ken, I would need to find something of interest about the environment or some kind of new information about any species for us to have conversations about. I also needed to get off the computer and get my clothes and car ready for the trip ahead. It was only two hundred miles but my old car was so beat up I had to baby it to insure its reliability. I constantly watched for leaks and had the oil changed. It ran great in the city, but out on the interstates where you had to drive fast it did worry me.

I had just gotten into Discovery.com when I had a voice request from Dodger. Oh I wanted to hear his voice so I accepted and the first thing he said just made me smile. "Hello, me love," he said and my heart melted. Oh I could have fallen in love with his voice alone but I was glad he was my friend. We talked for a few minutes and he said he had a poem for me and wanted to read it if I had time.

"Please read it," I said and he started.

"When I see a rose I think of you, in all its beauty, your image shines through. The thorns on its stem are just life's little bridges, and the delicate fragrance makes it that more delicious. So strong, yet so fragile, so small but easy to handle. When I see a rose, I think of you."

Silent for a second and grinning at my monitor, I told Dodger that it was the most beautiful thing I had ever heard. He chuckled and said, "Well, it's short but it came from me heart, me love."

We talked for almost another hour after that, but there was no mention about our playing and that was nice for a change. At least it made our friendship seem more real. If he had wanted to play, I think my feelings would have been hurt. Saying our good-byes until later, I was off again searching for a topic for the weekend adventure.

Chapter 10

Searching for a topic was not that hard to do and I quickly found what I was looking for and printed up some material. With that out of the way, I decided to surf through the chat rooms. Most of the time, I would check out the onscreen conversation and if nothing was happening there, I would move on. I had just entered a chat room in 50s Love when a female chatter started giving some poor guy hell for trying to cyber with her. Of course, monkey see, monkey do, and all the other ladies started typing on the screen about how they were ladies and did not cyber either. I sat back watching for a few minutes as they all congratulated themselves for being so grand. Since I was under a different alias, I typed in, "You ladies sound so precious and so sweet. But if the truth was known, I would be willing to bet that there is no one in this room who has not tried cybersex." Knowing that I had just thrown the shit into the fan, I laughed and surfed on to another room. Of course, I changed my alias again because I was getting PMs from pissed off women. I laughed so hard at each and every one of them as I deleted the PMs and moved on my way.

I was doing so good, feeling like a million until I surfed into a room and looked into the chatter's list and saw one of Ed's usernames. I froze in my chair. I wanted to leave but I could not move. Do I say hello or do I stay silent, I questioned myself, but he must have spotted me because he left the room. There I was, having a great time at the top of the world and with one glimpse of his name, I became a babbling idiot—a stupid twit. Now would be a good time to get offline and take a bubble bath, I thought. I shutdown all my programs but left the computer on standby. Right at that moment, I did not want to think, I only wanted the feel of the hot water soothing my body.

Setting the CD player, I filled the tub and added extra strawberry foaming liquid and watched the bubbles rise. Slipping into the silky, fragrant water, I leaned back, closed my eyes, and practiced the meditation breathing techniques I had learned from the Net. I opened my eyes later to find the water a bit chilly and glancing down, I saw my nipples almost firm. I pulled the plug, allowing some of the water to drain before turning on the hot full force until I felt warm again. I soaped up my sponge and started rubbing. I could not help noticing my breasts. My breasts still felt very sensitive when they were touched and I had often wondered if the feeling would disappear with age. I had never asked another older woman if the feelings ever went away. Maybe for each woman it was different. I just knew that it sure wasn't a chat room topic. I giggled at myself. I could not see myself asking another woman that question, on or offline. I could research on the Net, my window to the world.

That night, Jim called while I was in PM with another man and I had a great time flirting with both at the same time. Jim could hear me typing and got a wee bit mad, but I explained that I was talking to an upset girlfriend. Oh, I was wicked at times, grinning like a big cat with a fat bird in my mouth. While I was talking to Jim about going to Kentucky, I was also smooching and groping a man in PM. At least my staying busy kept my mind off the one I truly missed. My screen was exploding with PM messages and I hated when that happened. I mean, I am good but I cannot do more than two things at a time. My typing was hunt and peck, so I was slow. I told a few that I was on the telephone and the others, I just ignored. They would be there later so there was no sense in worrying about them now. I wish I could have fallen in love with him.

Going back into my chat rooms after Jim and I finished our call, I found Kate, who was just as bored as me by then. We played like children, laughing and cutting up. I was playing the part of a hawk again. We surfed through a few rooms talking to those we knew but staying in constant PM with each other. I was doing great, then out of thin air, Ed pops into the room and my heart fell to the floor. He saw my name and immediately typed on the screen that he was leaving. No, I thought, and double-clicked on his name, which brought up his PM window.

Typing fast I said, "No, do not go, please stay. I was leaving anyway." He typed back that he did not want any problems.

"There will be no problems," I answered, then left the room. God I wanted to run and hide somewhere. At times, I wanted to shut my computer down for about six months and never come online again. Nothing was the same for me but then again, maybe it was not supposed to be. Telling only Kate and Gwen that I was going to Kentucky for a few days, I decided to go to bed. My night was shot to hell now, so why prolong the agony?

The next morning, bright and early, I jumped online to check my e-mail boxes. Before long, I was locking everything up and Rebel and I were heading to Kentucky. The weather was chilly but the day was sunny. Hopefully, it was all going to be okay. Laughing and playing the radio very loud, I talked to Rebel while I drove like he was another person. Of course, he did not answer me but he would tilt his head from side to side making me laugh that much harder. I had to wonder what he was thinking. If he could have spoken to me, what would he have said? "Hey stupid, slow down, you humans are nuts, or do you wish you could lick your ass like me?"

Rebel had been my pal a long time. As I watched the countryside, I thought back to the first Rebel. I had gotten him when I was ten and he stayed by my side until he died eleven years later. He was all I had in the world as a child. He had been my friend, my playmate, and the only live thing I had to talk to until I married. When he died, I had just given birth to my second child and for me, it was a sad day. Granddad had just finished burying him when I got home that day and I stood over his grave crying for hours. It took me over twenty years to bring myself to get another dog. When I picked out this present Rebel, it was because he had the marking just as the first one did. I named him Rebel, of course, and he would stay with me until he died but after that there would never be another dog for me—I had made my mind up on that already.

The time passed quickly and soon I was pulling up in front of Ken's trailer.

His trailer was nestled off the main roads in the tall trees and it was a perfect place for a naturalist. It was also a place Rebel could run free without the fear of being run over by a car. The air seemed fresher and there was Ken smiling a great big smile as he met me in the driveway. He almost pulled me out of the car so he could hug me and we were laughing at each other. Ken was almost fifty now and I could not comprehend why he had chosen to live his life in solitude. Working with the animals and watching baseball games was his whole life now, but as much as I did not understand his reality, I respected him. It was his choice and whatever he did, he would always be my friend. Still, I hated to see him alone.

We moved my bag into the spare room then settled in the living room to talk. He had his beer and I had my Pepsi, so we were ready. Watching his face while he did most of the talking, I was captured by his intelligence. I knew that he had a brilliant mind and had been offered a place in the famous MENSA group many times. He had always said that he did not want to be bothered, that the path he had chosen was his choice. Talking with him was always interesting for me and he knew me like the back of his hand. Finally, after about two hours of his talking nonstop, he said that he needed to check on the snakes at the wildlife center and so we jumped into the truck and headed there.

I had worked at the wildlife center some years back but when my husband and I split up, I moved back home and rarely made it back anymore. My ex had been gone sometime from the area and this was my first visit back since then.

While Ken went into the snake room, I turned on the compound's speakers and the music poured out over the natural settings. Opening the back doors, I was amazed by how little the place had changed. Roaming down the paths and around the pens, I greeted all the wonderful birds and mammals. There were even some animals there that I had lovingly nursed and raised but like me, they looked old now. I went to each kennel talking and rubbing all my old friends. I felt like a child in a large playground. My friend Lupia, a timber wolf, gave out a loud, lonely howl without warning. It was an intense howl of loneliness and it made the hair on my head stand straight up. Making my way to her, I saw that she was running around her area, jumping at the fence and whining. Smiling I said out loud, "Oh my God, you remember me." I moved to the fence and rubbed her head, letting her lick me through the wire. I did not see Ken come up behind me but he spoke and it startled me.

"As I looked around here, I do not think any of the animals have forgotten you," he said, and then he said that if I wanted to go into the pen with Lupia, I could. The place was closed and he would stay close by in case Lupia decided I was lunch and we both giggled.

Going into the pen of an alpha female timber wolf was something not everyone can do. These wonderful creatures have more than a sixth sense. Lupia was very special though she had always been given extra attention by the men working because they could not put another wolf in with her. I played with her, rubbing her giant frame and letting her lick my face, then throwing her treats and watching her play. Sometimes, I wish I could have gone back to Kentucky to help here again. It saddened me that my heart and legs had failed me so when life finally could mean so much. I said my good-byes to Lupia and felt very sad leaving her there again.

Ken locked the place up and we headed back to the trailer. He talked all the way home and when we were settled back in, he grabbed his guitar and started to play some old, slow rock tunes. I leaned back listening for a while when he finally stopped in the middle of a song and looked at me.

"Okay, what is wrong?" he asked.

Yes, he knew me to well, I thought. He lightly played a Bob Dylan tune as I talked about everything that had happened and how my heart was broken and how stupid I had been. How my obsessive behavior had made me lose the one thing I wanted most. Ken never said a word, as I must have talked about an hour before finally taking a deep breath and leaning back to take a drink of my soda. As if he were analyzing everything I had just talked about, he took a long drink of beer and moved over to the couch beside me, laying his guitar to the side. He took my hand and started talking to me, looking directly in my eyes.

"I have known you for many years," he said softly, "and you are one of the worlds most loving, generous ladies that I have ever known. You are kind, considerate, and you're almost as smart as me," he added, flashing me a giant, humorous smile. "I see nothing obsessive about fighting for what you believe in. You have always fought for what you hold to be true and there is no shame in that. Do you not understand yourself, honey?" he said. I must have had looked confused so he continued. "You were given unbelievable, shitty odds as a child but you fought against the hardships. You educated yourself without the benefit of any help or guidance from anyone. There was no one in your life to help you grow to your fullest potential and yet you and only you made it through, carrying your battle scars not as badges of defeat, but badges of honor. You have more compassion for life in your in little finger than most people ever experience in an entire lifetime. Never be ashamed that you are a warrior because without people like you, this earth would be one miserable, disgusting place. You, my dear, are one of God's most precious wonders and I will always be proud we are friends." He turned and hugged me very tightly. When he let go he smiled and said, "Okay, do you want to teach me about my computer?" Laughing, I agreed as we dusted off the desk and turned the old computer on to get started.

We stayed up until the early morning hours until neither of us could see the computer screen any longer. Shutting everything down, Ken went his way and I headed to my bed. It was the first full eight hours of sleep I had had in many months.

The next morning when I woke up, there was coffee ready and a note saying that he had to run and check the snakes again. I grabbed my coffee and went onto the porch. Although it was chilly, it was good to sit there in the mist of the trees just quietly thinking and relaxing. I thought about what Ken had said and I knew him too well, too. If he thought something was not right, he would tell you in a heartbeat, that was just the kind of person he was. He never minced words about anything; he was straight and to the point in his conclusions. Ken was a great man and a dear friend.

Catching a cold breeze up my robe, I headed back inside and decided to go online to check my e-mail. I somehow felt very peaceful about myself. There was no rushing about or frantic behavior and I felt like I had when I was younger—strong and able to do whatever I needed to do. Taking time, I analyzed where I was at now in my life and what needed to be done to make improvements. First of all, I had to let my heart close where Ed was concerned—that part of my life had to be truly over. I had done my best and it would always be his loss. My heart would have a little place with his name etched in, but that would be it. Second, I needed to develop another positive side. I realized that what I had said and done in the past I could not undo, and that I could only build from this day forward. Still, there was something missing and I could not put my finger on it yet. I wondered how to handle all my sexual escapades online and if I wanted these to end. They were fun and they did not hurt anyone.

For the next few days I did research online and Ken silently watched and listened as I explain the fundamentals of basic computer operations. Every so often, I would go outside to play with Rebel and to take in some more fresh air. I picked up a blanket off my bed the last time I went back outside and laid in the hammock. Everything was covered but my face as I watched the leaves in the trees above me waving in the breeze. Something that I used to have had disappeared from within me lately, and that was my spiritual side. Maybe that was the reason I had become so unbalanced, I thought, as I drifted off into a peaceful sleep again.

I must have slept a good while and when I woke, rain was hitting me in the face. Trying to get myself out of the hammock was impossible. The cold had stiffened my legs and they would not move. I could see Ken's truck in the drive and knew he was home but I was not sure he would hear me if I screamed. No matter how hard I worked, I could not rub my legs enough to get the feeling back. My toes would not wiggle, and my hips would not move. Oh good Lord, I thought, another fine mess I have gotten myself into. I began shouting for help and I must have screamed about a dozen times before Ken opened the trailer door and saw me struggling. Rushing to me, he tired to lift me but he was not strong enough. He then started rubbing my legs and feet but nothing was working. Ken was ready to call an ambulance when a big red truck pulled up in the driveway. I was so embarrassed that I lay back down covering my head with the now soaked blanket. I was taken by surprise when two big arms slid under me and I was being lifted from my doom.

"Excuse me, miss, but you seem to need some help," the shadow said.

"Yes I do, thank you so much," I replied. It was dark and all I could see was a large figure wearing a hunting hat, but I was saved and very grateful. Ken had filled the tub with warm water and had the coffee brewing. The stranger carrying me walked down the hallway to the bathroom, then lowered me clothes and all into the water. He turned, quickly leaving the room before I could see his face.

In the tub, I was able to work with my legs until the feeling started coming back. Ken brought me a cup of hot coffee. When he set the cup on the side of the tub, I whispered to him, "Who is that man?"

"A friend from down the road that was coming over to watch the baseball game with me," Ken said. I asked Ken to bring my bag so I could get dry clothes, feeling like a moron for this happening. I heard the man ask Ken as he got my bag, "What is wrong with her?"

"She has Arteriolosclerosis in her legs. This does not happen often but the cold got to her when she fell asleep outside," Ken spoke softly.

I finally got the feeling back in my legs and feet, then dressed quickly and took care of the wet clothes. I dreaded going back into the living room to face the stranger, but I needed to thank him personally for saving me. I walked into the living room, moving very slowly. Ken had prepared a place on the couch so I could be comfortable and put my feet up. Laughing about what a nitwit I was, I looked up into the hero's face for the first time and my heart stopped. He was the most handsome man I had ever seen. He was over six feet tall, with sandy-colored hair, and the deepest hazel eyes in the world. Both men stood up to help me get settled on the couch, then Ken introduced me to Mike. With my best smile, I greeted Mike, thanking him for coming to my assistance. I wondered if he could hear my heart pounding in my chest. Laughing to myself, I thought, calm down, girl.

Ken turned the volume down on the television and asked me what topic I had brought this time. Smiling at him I said, "Sharks." As Ken began telling Mike and I all about sharks, we became a captive audience but a willing one, too. My eyes kept drifting back to Mike as Ken spoke. Mike was a big, striking man with a deep laugh and it wasn't long until we were talking like old friends.

The time flew by and Mike looking at his watch saying he had to go. I was sad to see him leave but I understood as he rose from the chair to go.

"I hope I get a chance to see you again," he said just before closing the door.

Turning to Ken I asked, "Was Mike blushing when he left, Ken?"

"Yes, I believe he was," Ken said. "You must have made an impression with him," and then he went back to the baseball final scores. He had made an impression on me as well, I thought, thinking of his laughter again.

Turning on Ken's computer, I let my mind go back into the spoils of the Internet telling Ken that I would have to go home tomorrow. He nodded and said for me to come back anytime or I could just move in with him. We could share the rent that way and we would not be lonely. Smiling at him, I said I would think about it and the truth was, I had already thought about it many times but now he had Mike to give me extra incentive.

As I surfed the Net, I kept thinking about Mike and wondered if he was good in bed. Something about the way a man moves can tell you so many things.

Ken's computer was not as good as mine, but I did manage to get into the chat rooms and find my home girls. After setting up a voice conference with Kate and Gwen, I told them about my daily adventures and that I would be home by tomorrow afternoon. We were laughing and cutting up in voice, exchanging all the daily gossip when Kate said, "Look who just came in." Glancing into the room, I saw Ed. I can't say that my heart stopped, but I felt alone again in the world. Returning to voice, I jokingly said, "Maybe he will find what he is looking for soon. He's a whore looking for whores." I laughed. We talked a few minutes longer while I kept a vigil on the main chat room. Saying I would catch them later, I turned off the computer and decided it was bedtime. Saying goodnight to Ken, he turned to me, asking what was wrong. Before I could say anymore he spoke up.

"You saw him," he stated. "I can tell by the look on your face." He hugged me and I headed to my room.

My sleep was restless as I dreamed of talking to Ed and asking him over again, why. My answer never came and all I heard was him singing that stupid little song he used to sing to me on a video program. *You Are My Sunshine*. How I had grown to hate that song.

The next morning, Ken wanted to show me a spot by the river he thought would be great for fishing when spring came again. Driving down the narrow lane in my old car, I saw a beautiful grove of trees as we pulled up to park. "Wow," was all I could say as we stood on the banks of the Tennessee River. According to Ken, these trees were not common to this area and must have been planted by man and, judging from their size, were thought to be at least a hundred years old. Standing in the middle of these trees, I sensed a strong power of wellbeing. I had read an article on the Internet about groves being planted in the aid of producing high-energy spots and by golly, now I believed it. I was amazed that I was actually seeing one. There were many thoughts running through my head as we left the spot and headed back to my car.

After Rebel and my things were loaded in the car, I asked Ken to tell Mike to e-mail me and added that I wanted to do a ritual in the grove when the weather warmed a bit. Ken's eyes opened wide but all he said was that he would help me anytime I was ready. Looking at me he asked, "What kind of ritual?" I told him I was not sure yet but I knew I needed to do something.

Saying good-bye, I headed home, my mind filled with all sorts of unusual thoughts of rituals, computers, and new people in my life. I laughed at myself wondering if I could talk NASA into shipping me off on the next space shuttle to Mars or some other planet far from this one. I was a dreamer but then, dreams were all I seemed to have at this point in my life.

The two-hour drive went fast and soon we were home again with Rebel looking disgusted. I must have looked disgusted to him, too, because I hated to put him on his chain but in the city, there were no choices in the matter. I quickly unpacked everything, made some coffee, and logged online.

Opening my mailboxes and checking e-mail, I had a couple of new ones with names I did not know. Giving it some thought, I recalled that they were from Yahoo personal ads that I had answered. After reading them both, I closed one out and began typing to Earl, who sounded nice. It was the usual exchange of basic information letter but that was the normal procedure at this point. After sending the e-mail, I noticed my Dodger had sent a card and so had Jim. Jim wanted to know where I had gone since he had tried calling a few times, but got no answer. Rodeoman left a note saying that he wanted to talk with me again but I was cutting him loose. That man was getting too intense for me. Still, the one e-mail I had hoped for was not there.

I decided to surf a bit, making the pagan site I had bookmarked my first stop. As I read further into the article, I discovered more and more things dealing with the Native American cultures. Native American cultures were my main interest and I spent hours reading every article I could find on the subject.

Feeling restless, I surfed the chat rooms playing my songs and watching the rooms and the people playing. My homeroom had so many new people now and it was not fun anymore. I decided to make my own room and to listen to my songs and work in my journal. Sitting with my headset on, I heard a song start playing and, looking up from my book, I saw Ed. The song he played was a Faith Hill song and by the time I got hello typed, he was gone. I sat there listening to the song, *Just To Hear You Say You Love Me*. What is going on with that man? I thought. Did he want me to follow him? Well, he was out of luck this day since I was through following and worrying over him. Before I could change rooms, he suddenly reappeared. I laughed because now I was sure he was nuts. Ed started proclaiming how much he had missed me and how lonely he had been without me. All the time I was thinking, this man must think I am extremely stupid, but once again, I wanted to see how far he would continue with his charade. "Ed, do you have your headset on?" I asked. He answered, "Yes," adding a smiley face. Turning up the volume on voice chat, I spoke softly asking him to slip his jeans down and he responded with a, "Yes, dear." It was difficult not to giggle out loud but I was super cool telling him I had my little black nightgown on and was laying on the couch waiting for him. He talked about kissing my face and rubbing my body with his large hands and how soft my body felt to him.

"Oh yes, darling, I love it when you touch me. I can hear you breathing harder," I said and he laughed and said, "That is not all that is getting hard." We continued, lost in the fantasy. He nuzzled my neck then started kissing me telling me how wonderful I tasted as I laid back and enjoyed his voice. I told him I had been dreaming of holding him and touching him. I wanted to feel his body naked next to mine. I was on a roll so I kept talking about his hands cupping my breasts, sucking them tenderly.

"My hands work at delighting your skin," I said. "I'm running my nails up and down your back as you move over me. Your legs entwine with mine as we move together like the perfect dance couple." My breathing and his were almost right to go for the kill, I thought. "You talk of kissing my legs, moving up to my inner thighs as I stroke your dick slow and firm." In my headset, I heard him moan and I smiled.

"Stop," I said into my microphone. "I now move over your body, letting my breasts brush your chest as I kiss your face and your neck, running my hot mouth down your stomach and allowing my hands to dance across your skin, tickling you a bit but demanding your full attention. I press my mouth into your stomach, as my hands play lower, feeling your full hardness. I move my body up further, opening my legs to straddle you. I lower myself down very easy as you move and moan with me. Slowly, I let you in deeper and deeper, rocking my body on yours. Your hands reach up to play with my breasts as I place my hands on your stomach so I can steady my body as I ease further down on you. I hear your breathing getting faster as I rock and move my body up and down, feeling you move with me. Now, love, you finish this for me," I whispered into the microphone.

"I pulled you off me and lay you down as I mount your body, feeling like superman, taking charge. I insert myself into your hot body, feeling the tightness surround me. I

122

hold my breath a second as I resist pouring out into you. You're so hot inside, so very tight, I could explode now as every sense I have tells me that you are mine and mine alone. Moving myself in and out, I feel the sensations run like lightening from my head to my toes. I thrust deeper and harder each time as you move your body in rhythm with me. Your hands are holding, grasping my hips, pulling me, and pushing me and we start moving even faster. Your soft moans tell me that all is right and wonderful. Oh, darling you are so wonderful, I cannot help loving you. My body is out of control, I cannot think about anything but the feelings getting stronger. I want to possess you, to devour you as each second is filled with total passion. There is no way I can stop now, no way I can hold back any longer. The time is now, I am letting go with all of me in you. One last thrust and my body erupts as I hear you following my lead."

Silence followed and I could not hear anything but his jagged breathing in my headset. He groaned loudly from the pleasure that was obtained at the moment. I smiled knowing that this had been a successful meeting of both minds and bodies. Having complete love physically and mentally is the greatest part of life one can have and my life was complete for now. I felt no shame, only love and admiration for this man. Too bad it had not been in person, I thought, getting myself back together after releasing the moment's passions.

I reached for a cigarette trying to get my composure back. Thinking hard for something else to say to you at the moment, I could only shake from what had just happened. It was impossible to think now. Finally, speaking again, I said, "You are simply great, my dear, and I think I will go take a nap now. I do hope we can chat again later," and he said, "That would be fine." Laughing, he also said that he needed a nap and we both laughed as we disconnected the voice chat and went our separate ways for the time being.

Having every intention of taking a nap, I thought I would check my e-mail first and much to my surprise, there was an e-mail from Ken. He had written wanting to know if it was okay for Mike to e-mail me. Ken said, "Ever since you've been gone, that man has been asking about you and wanted to keep in touch." I did not know Mike had a computer, but he could e-mail me anytime he wanted, I told Ken. I suggested that he download the Yahoo Messenger, then we could voice chat. Pushing the send button, I wondered why Ken had not used the messenger to talk to me, but he was a genius with everything except computers. I had a reply from Earl, which was a surprise coming back so fast. Instead of answering his e-mail, I thought I would take a break and get things done around the house.

Turning on my stereo, the housework went very smoothly. Dancing around getting caught up in my music, I forgot about the computer for a few minutes. It did not take long to put everything in its place and dust the few pieces of furniture I had. After that, I settled down to watch a movie and write in my journal. During the movie, *Braveheart*, I must have dozed off but I did not miss any of it because I had seen it thirty times in the last year.

I turned my computer back on and, as was my habit, I opened my e-mails first thing. Glancing in, I saw a strange name and opened it up reading it out loud before I started laughing my ass off. It was a letter from one of Larry's old girlfriends. It was the lady he'd literally dumped when he and I became an online couple. She was now informing me that I should stop sending Larry e-mails. That he never read anything and that every time I sent any mail, he forwarded it to her. She said that she and Larry had a very special relationship and that I did not have a clue as to what Larry was really like. In the letter, she wished me luck and happiness and that I should leave Larry alone because he would never come back to me. She also added that she had sent him a copy of the letter and he'd approved it being sent to me.

Well now, "never" was a long time, I thought, as I continue to laugh. This silly woman was so stupid that she did not have a clue that Larry used several online aliases and of course, Ed was one. Furthermore, I could tell from the subject head that the e-mail she was talking about was not the last one I had sent Larry, and what kind of real man lets a woman do his talking for him? If Larry/Ed wanted me to stop the e-mails from coming in, why didn't he put a block on my name? Let's face facts, the man wanted me to continue to send him mail. I knew he was somewhat computer stupid, but even the dumbest person knows how to block unwanted e-mails.

Yeah right, I thought, still laughing at her. The poor idiot did not have a clue that Ed was online more than she knew. I sat back a minute and thought over this situation. Was Ed trying to get me mad enough so I would tell her everything I knew about him? Had he dug such a deep hole with her that he wanted a way out? There was always the possibility that he didn't love me, but I just couldn't see that happening. This Larry had told me out of his own mouth sometime that he could never care about this lady, but something was sure up here. I did not know what the deal with him was; this only made it more interesting for me. "Leave him alone," she'd written. There was not a snowball's chance in hell that I would leave him alone now. Would he forget me? Nah, he would never forget me. I knew I would haunt his mind for many years to come. Even if I did back off, I still felt it would never be over with him. No matter what this nasty mouth bitch said. Maybe it was fate or maybe we were soulmates, I was not sure. This fired me up more, but I was going to play it cool for now. Shaking my head, I thought, I may not leave him alone but I wanted a man with a backbone and he was a wimpy, spineless bastard. The evidence was plain and simple—the man was a coward. I wish I could forget, maybe in time I could, but it would be my choice and my choice alone. I did not want to give up the dream yet. As long as Larry kept coming to me under an alias, why should I stop? Besides, he was so good at turning my world upside down.

Hitting the reply button, I started typing. "I wish you and Larry the best of luck and I know that you honestly deserve each other. Best Wishes, Veronica." That was short, sweet, and very nice of me and should satisfy Ms. Piggy for the time being. Whatever his game was, I was not going to play it with her. If he wanted me to leave him alone, then he was going to have to tell me like a man. He never gave me any indications that he wanted me to stop either. Still, I would not be used to help him on this one and besides, I had him almost anytime I wanted. With all this new information, I was content to play with him knowing the truth.

Moving on to my next e-mail, I saw another strange name. It was a nice e-mail from Mike who said it had been very nice to meet me at Ken's. He said that he had not been online long and did not know a soul to e-mail or talk to. I wrote him a nice e-mail back and sent him directions on how to set up the Yahoo Messenger for easy and better talks. I thanked him again and said I was in his debt for saving me from the cold and I hoped that we could become friends. I signed it and pushed the send button.

Now this could become very interesting, I thought. I could always use another friend, but the telephone ringing interrupted my train of thought. Answering it, I heard my wonderful Jim on the other end and that made this day complete. Jim and I talked for an hour and that was a short conversation for us, but he was at work and busy that night. I kept asking when was he going to retire and he would laugh and answer, someday. Talking with him was always refreshing. Although he had been in California for many years, he still carried his Boston accent. Sometimes I could catch it in his voice and would giggle at him and he had no clue why. We talked about everything in our allotted time and then he was off, promising to call soon.

Although I rated Jim as a very dear and loving friend, I could not fall in love with him enough to say yes to marrying him. With one simple yes, I could move to California and never have to worry about another thing as long as I lived. Sometimes I thought I could marry him and deal with the love part later, and I did love him, but not with all my heart. Boy was I sap or what, I thought to myself. Yet, tomorrow was another day.

I was up very early the next day and online as always. I wanted to continue my research into the mystical paths of the Native Americans. There was lots of printing and studying to do. I had found a ritual dating back hundreds of years that my ancestors had performed for cleansing the soul and for bringing one's true destiny forth. Even though I was a bit afraid of what my destiny was going to hold, I was ready to move on with my life no matter what it held in store for me. I kept asking myself, did I believe in all this hocus pocus stuff? My answer was simple, what could it hurt? Still, there were many things I had to gather for this act and a lot more I needed to learn. It was so like me in these cases to make everything as right as possible. What if it did work and I screwed it up not knowing the correct timing or words? I studied harder and surfed more into the realms of the religions I once thought were fantasy or just plain folklore. While organizing all my printed pages into a folder, my Yahoo Pager beeped—it was Mike.

"I am glad you made it and congratulations," I said. He was new to the Internet and he was not sure what he was doing, so I suggested that he open up the voice chat and I would help him get everything set up correctly. Once he opened voice chat, he was like a child at Christmas with a new toy. He talked and talked and I laughed at his Net innocence. The one thing that impressed me the most was the way he laughed. I loved to hear another person's laughter and he laughed about everything.

We talked for two hours about everything we could think of. He had never been married but had been involved in a couple of long term relationships. He had no children. He had dedicated himself to working with wildlife and doing school programs for children. His job was taking live mammals into the schools and teaching children about them and their environment. A man who had passion for life and nature was a strong combination for enjoying life, I felt. I sincerely hoped that we would become good friends because everything he talked about was what I enjoyed, too. My wildlife had consisted mostly of the silly people online. Mike and I talked for thirty more minutes, then I needed to go. We said good-bye and I left the Internet to lay down. The new heart medicines my doctor prescribed for me were making me very weak at times. I did not like that feeling of weakness and fought it, but to no avail. I was advised to take a nap or to at least lie down on the couch for an hour or so daily.

Earlier that day while in my home chat room, a lady chatter who I knew had ask me about writing a short paragraph on the Trail of Tears; she knew of my Cherokee background. She was putting together a web site on native Americans and I felt honored that she'd asked me. Closing my eyes, I placed myself on that trail with my people and this is what I wrote.

The Trail Where They Cried

I stand up straight and look ahead—there are all my people being driven like the buffalo. Over a land with no paths or roads, through the rivers and the mountains. Most have no coverings on their feet and very little clothing to protect from the harshness of the weather at times. The Mother Earth has cried with us so many tears that the waters have risen to make us stop, but the army men push us still. My feet hurt and my legs ache as we have walked many days with very little rest. We have very few wagons for the old ones and the few things we were allowed to bring. The old ones are dying and the young ones cry for there is not much food. Our braves are only allowed to hunt for short periods and the Army men don't give us much food, yet, they call *us* savages.

I don't understand why we must go away from our homes. We are a peaceful people with our farms and animals. Yet, the "wise" ones say this is the best for our people. At night I listen to the men of the tribe speak of lying and mistrust where the white eyes are concerned. In the dark, I see my sisters being taken away by the soldiers and when they come back, I see the shame in their eyes. We turn our heads, not for their shame, but if we don't accept this terrible truth then it has not happened.

The children are told we must live in harmony and peace and the place we go to is a better land. Still, their bellies ache from the food, which has bugs and sometimes the supply wagons are very late. Sometimes when they get here, they are even empty.

The white eyes give our men a drink that makes them act silly and they dance at night and sometimes cry, too. I had never seen one of my brothers cry until this journey started. There is very little laughter here and every day, my sisters and I pray as we walk. Sometimes we sing to the Mother Earth for protection and guidance and to keep us strong until we reach this land we walk towards. I have never known being hungry like this and my body bleeds from the cuts on my feet. There is no complaining about these conditions for fear of being hurt. I pray that the child I carry will wait until we come to this new land before entering the world and when he is old enough, I will tell him of this journey. I will tell him his people were strong enough to endure this thing that has been forced on us. I will tell him never to forget—The Trail Where They Cried...

—written by Whiterose, a.k.a. LadyHawk

It was not hard for me write down my feelings about the Trail of Tears. To me, it was another part of the tragic past of our country's history. It was an important part of our history along with black slavery and the witch-hunts of Salem. Except, my people's suffering and discrimination lasted longer than these other tragedies did. When I learned I had part Cherokee blood, the Internet opened the window of unlimited resources to learn about my Cherokee heritage.

Whiterose is an Indian name that I'd proudly taken after I understood my ancestry. The name came from a legend about a Cherokee Indian Princess who had befriended a family of white settlers traveling through the wilderness of the Appalachian Mountains. The legend goes like this: Whiterose was traveling to another village when she discovered a family of white settlers. They had been traveling over the mountain when a freak winter storm stranded them. The animals were dead and the family was freezing. The wood of the wagon was so wet it would not burn. Giving the family the food she had, she headed back to her village for help. She was riding hard, she was cold, and she knew the family had to be rescued before another storm hit. The clouds were coming in fast and the wind had picked up and she lowered herself closer to the horse while he galloped like a true warrior. Holding on with all her strength, she glanced up long enough to see her village ahead. Feeling relief, she pushed herself closer to the neck of her horse for the warmth when he blindly stepped into a hole. Whiterose was thrown against the ground so hard that she died, but not until some of the village hunters found her and the horse. Lying in the fresh fallen snow, she told them in a weak, muffled voice of the settlers on the mountain before closing her eyes. The settlers were saved the next morning.

Thinking of Whiterose, I knew I had a fighting spirit as she had.

I was up online again. Was I going to war? I thought as my computer connected. No war, just wisdom, and from now on, I would try to use the wisdom I had been given.

I took a deep breath, then entered the chat rooms to play. Breaking my habit of checking e-mail first, I bounced in and out of each room. I made no bones that I did not care about finding love in the chat rooms any longer. At that time, I felt that there was no hope for me in finding the love I wanted or needed, but tomorrow is not carved in stone for any of us. However, I did enjoy flirting with the homeboys because they knew I was just playing. The few men I had gotten close to in chat seemed to respect me and that was enough for me. Yet something was still bothering me, I had always brushed it off, but the feelings were now intensifying inside me each day.

Chapter 11

I was beginning to relax in my home chat room when my Yahoo pager beeped that I had a message. Great, it was Mike and I was ready to talk with someone as long as I did not have to answer questions like, do you like sex? Or, what do you do for fun? That last one always made me laugh and normally I would answer with some off the wall bullshit. I could be as nutty as the jerk and still come out ahead each time. He invited me into voice chat, which was more to my liking because I hated typing. We were talking again like long lost buddies and he seemed to have a gentleness about him when he spoke but I knew for a fact that he was a very big and strong man. Captivated by his voice, I could hear his kind soul in his voice, which spoke of the Mother Earth and her creatures. I listened intensely as he spoke with knowledge about pollution problems and not enough natural resources for future generations. I was simply amazed to have a nice conversation about something that really mattered and made sense.

It hit home with me when I thought back to when I used to care about these things, too, before I became a chat room groupie or Internet junkie. My thoughts were making me feel restless. Mike and I continued talking for two more hours. His laughter was like medicine for me; a deep rich laugh and I laughed right along with him. When all the stories were told and the laughter slowed down, he asked if he could drive down to Jackson to see me and to have dinner. I knew Ken would have informed me if something had been wrong with Mike. So I said yes to next Friday evening. Closing out voice chat, I felt very good for a change. After accepting the dinner date, I realized that, over the last few years, I had glued myself to this computer, shutting the door to the real world outside. In all fairness, there were many good things I'd learned from Net, I thought, trying not to think only of the negative things but at the positive things as well.

I went surfing into my pagan sites to investigate ancient text materials and religions from all over the world. For the rest of the evening, I did not go into the chat rooms. Printing many of the web pages as I found them, I surfed until my eyes could not see the screen and then I paged Katie and Gwen and said my good nights.

The next morning over coffee, I began writing on a piece of paper. On one side, I wrote everything I did not like in my life. On the other side, I wrote the things I did like in my life. I had come full circle, as I had heard said before, or maybe I was going through a mid-life crisis. Could it be? I panicked at the thought of dying so young. Whatever the reason, I knew I finally had to stop procrastinating about a change and make one happen. Now I had to figure out how. Looking at the material side of my life, I did not have much to start a new life with and physically, I could not work, but as my mom had always said, "Where there is a will, there is a way."

Getting another cup of coffee, I decided that I needed to work out things emotionally to take care of the things I could right now and to let the past go and dream about the future in realistic terms.

Getting dressed, I went out outside and played with Rebel. When he finally grew tired, I walked around in the yard taking in deep breaths. Sniffing the slight wind that surrounded me, I knew that rain was coming. I stayed outside a little longer, then went back inside to make fresh coffee. My mind was working overtime as I seemed to pace across the apartment. From now on, I would try to keep a balance in my life—a check on reality. There was nothing wrong with fantasy as long as one remembered it was just that. Besides, the weather was getting very cold and my legs were acting up again, which made it hard for me to be outside or driving. It was either television or the computer for me now, and I had seen enough movies in the last two weeks to last me another three years but hopefully, I would not be the Internet Junkie I had once been.

Shutting down the computer, I went out to buy supplies for the next week. The weather said snow was heading in and I had Mike coming to dinner on Friday—if he could make it due to the weather. With the shopping I had to do, I would be ready for anything.

After dinner that night, I put everything I would need around my computer and went online. Checking e-mail first and making my replies took several minutes, then I went into my home chat room. As soon as my name hit the screen, everyone was sending hugs and good to see yous, back messages, plus many wonderful little smiley faces. I felt very good about being back as I was asked to play a song. I played a Blues Brothers tune, *Minnie the Moocher*, and I could play the big bird again and thump the keyboard while shaking my tail feathers. I laughed at the antics of the ladies playing like they were dancing and wiggling around the room also. Kate and Gwen were there and it was like old home week. Everything was great and we laughed and played into the night having tons of fun.

While watching the screen, I saw my name put on the screen in front of the whole room and it said, "LadyHawk you are something else and you have made me smile for the first time in a long time." I looked at the user's name and saw that it was a new one, dwayne2u. Nothing was familiar about the name and I replied on the chat screen, "Well thank you. I am glad I could make you smile," adding a smiley face at the end of my sentence. The whole room suddenly become quiet and I knew they were reading what Wayne and I were typing to each other. He asked if he could PM me and I replied that I was going offline and perhaps we could another time. Sending me a smiley face, he flirted with me by saying that he could not wait to talk to me and I blushed in front of my monitor. I was not trying to be rude but my legs were hurting and my eyes were burning. Besides, I was looking forward to the pow-wow conference with Kate and Gwen that we had planned as soon as we left the chat room. It was to start in about thirty seconds and I knew it would last at least another hour while we talked about all the gossip. All in all, it was a very good night and I felt good my first night back. All was well with the world.

The next morning, I was up at dawn. That was unusual for me, I had always been a night-shift person but my days and night were messed up. Since it was very cold outside, it took a couple of hours to get my legs working again, but then I turned on the computer and away I went to check my mail. I was truly amazed at all the people from chat who had seen me online the night before and wrote to let me know how good it was to see me again. There were many cards from Jim and Mike and I sat there smiling as I read each word from them. Just reading the e-mails took a couple of hours and writing replies took time, so it was almost noon before I had everything done with my mail. It was time to open up my other programs and see what waited there. Opening ICQ, I had tons of messages from the friends on my list. There were at least two dozen web sites that had been sent to me for viewing, which was so overwhelming that I had to save some to view later.

Opening up my Yahoo Messenger, I had lights blinking everywhere and loads of offline messages left for me. The first thing I noticed was that Wayne had asked permission to be added to my friends list. Oh what the heck, I thought, there is nothing wrong with getting to know new people. Jim, bless his heart, even with all his calls, sent words of encouragement with a promise to call later that night. Mike left plenty of messages saying that he was looking forward to seeing me on the weekend and I knew I was looking forward to seeing him again, too. Looking at the calendar, I had only one day to cook and get the house ready for him. Luckily I had bought all my supplies and had enough food in the house to feed a small army.

I decided I would not surf the chat rooms because I knew there was always a chance of running into people I did not want to see. Kate and Gwen both promised to be there if I got into trouble and that meant if Ed showed up I would just ignore him and talk to them until it was time to leave or until he left. Even though I felt stronger where he was concerned, I could never be sure how I would act when he came into the room. If I was lucky, I would never see his name again, but I had never been lucky, so I knew a time would come when I would be faced with that situation. Honestly, I did feel like I could handle it but I could not be sure. I was only human after all.

Closing down my computer, I wanted to watch some wildlife on the tube and Lou was due over to take me out to lunch. I was into a mild exercise program trying to get my legs into better shape. With all my efforts to build myself up physically, I was doing better with my mental outlook. My heart was still giving me trouble at times and the pain in my legs increased, but I refused to take pain medications again. When the pain got to be too much, I would lie down until it eased up.

By the time Lou came over, I felt better and was dressed and ready. We went to our favorite restaurant and sat at the table talking and laughing as we had done for years. Lou was great for encouraging me, too. She knew all about me—all the troubles I had seen over the ten years—not only physical problems, but she had been there with me when my marriage had broken up. In fact, she had firsthand knowledge of the bullshit I had gone through with that marriage. Frank had been a bartender in her family bar until she'd finally had to let him go because of his temper. That was where we had met in the beginning and become friends. She even came to me with the problems she'd had with him at work. I knew from the way that he acted at home, screaming and carrying on, that she would have to fire him and I could not blame her. I knew he was running customers away with his attitude. It had been a very bad time in my life and all I wanted was never to think of it again. Today, if someone screamed around me, I would leave. I could not bare it. I had no trouble forgetting Frank after the thirteen years of hell he'd put me through. I always seemed to have a problem picking the wrong men in my life and Lou and I laughed about it over our lunch.

The cold weather was making the arthritis flare up and my legs were not working well at all. After Lou dropped me off at home, I got on the computer to study my Cherokee pages. I could sit in front of the screen with my feet propped up and surf. My house was clean and the roast was in the crockpot slowly cooking. Everything was going well and I was moving right along with my language lessons when I heard a ding and my Dodger man was in my PM. Siting there a second, I read the PM. He wanted to know how I had been and where I had been. I realized that to him, I did not matter at all. All he wanted was the sexual release he had grown accustomed to when we chatted and in a way, I could not hold that against him. I had basically done the same thing to him because I knew nothing about him or his life. It had been a mutual pleasure seeking act on both parts but still I did not want to backtrack, I wanted to move on.

I chatted with him for a few minutes then told him I was busy and that I would catch him later. He did not seem pleased with that, but there was nothing he could do.

As he was saying good-bye, I received a PM from Wayne inviting me to voice chat and I accepted. Wayne and I talked for hours telling each other about ourselves and laughing about things that went on in the chat rooms. I actually had a great time talking to him and he had a great voice. With him, I felt the old sensations of lust warming my soul again as I laughed to myself. Though I wanted my life to change and felt like I had, I was still the same sexual and passionate person I had always been. Wayne and I continued to talk on and off most of the evening but I did not proceed to push the sexual part at that time. Everything within a time frame, I thought to myself. This was not the right time for us. It was time to meet my friends in our home chat room to play. I knew myself well enough to know that if I did not pull out of this conversation soon, I would be looking for batteries for my toy. Laughing out loud at myself, I bid my farewells and headed offline. There were things I still needed for all plans and so many store's to check out.

It was almost dark when I got home from shopping and I took care of Rebel. I looked at the computer as I worked around the house but there were other things that needed to be done. I gathered up all the information I had from the Net and read carefully for a few hours thinking and planning my next actions. Everything I had discovered about rituals seemed harmless and I was interested in seeing if anything special would happen. The fact that the ritual called for a psychotropic drug to be used intrigued me, too. Of course, all I could risk doing was smoking a joint of marijuana, which I had bought off of a neighborhood teenager. It had been many years since I had thought about pot but it was part of the ritual so I felt that it was okay.

Setting up candles all over the living room, I then placed white sage incense in the special burners. I had drawn ancient symbols on virgin paper, then placed them on the walls in all four directions and above and below. When the living room was set up to all specifications, a cleansing bath was next. Playing Native sounds in the CD player, I prepared my body. The bath water had pure rose oil and sea salt in it, and I had special incantations to say as I sat there pouring the water over my body. From the bathroom window I could see the full moon rising, which was all part of the plan. When I felt relaxed enough and had said all of the incantations, I moved my naked body into the living room.

The candles were burning brightly and the white sage filled the air. The Native sounds that also filled the air were turned up loudly and I could feel each drum beat as if they were my own heart beating. The coffee table was a mass of candles burning and things gathered from the Mother Earth. Leaves and acorns, a small bundle of straw, a small pile of stones, and a goblet of rain water from the last rainy day. Everything looked ready and I began.

Walking around the room carrying the white sage and saying the blessings, I moved to the couch. Picking up the joint, I continued saying the blessings over and over again as I smoked. Placing myself in a meditation position, I took a few deep breaths and called to the ancient Old Ones to lead me on my journey.

What happened next was the most exciting thing I have ever experienced in my entire life. I closed my eyes, I could feel myself drifting upward, and I could see myself leaving home and soaring through the heavens, touching the stars. The feeling was almost orgasmic as I felt myself being pulled down into another time to a place of lush meadows and volcanoes in the background. I had entered a time at the beginning of time it seemed. I was in a small valley surrounded by mountains and a thick forest where everything was green and pure.

There seemed to be mystical chants softly echoing around me. I wondered where I was and what I was doing here when, from behind, a giant black wolf spoke to me saying in a low voice, to watch and listen. There was no fear in me. Everything I saw and heard was very natural to me for some reason. I stood very still as I suddenly saw a figure walking toward me from out of the forest. I had no feelings of fear, only one of amazement as I watched the being come closer. His body looked to be very old; his walk was slow, and he was covered with thick, reddish hair. He had only a piece of animal skin tied around his waist and, as he came nearer, I stood very still in front of him. When he was close enough I asked, "Why am I here?" and he looked at me saying, "You have always been here." Looking over his shoulder, then back to me he said, "I have been with you always." When he had finished speaking, a sound of metal being struck sounded throughout the small valley. He turned and started walking away from me. While I stood there, I witnessed many figures of the same type coming out of the forest and heading for the same cave-like opening at the other end of the valley. Out of curiosity, I wanted to ask more questions but I knew he was too far away to hear.

Black Wolf appeared at my side saying, "Get on my back and hold on." I obeyed without question. We began moving fast and I buried my face in the fur on his neck, but I felt happy. I was part of everything and understood nothing, but I felt free and lighter than air. My hands were fists as the black spirit wolf soared back to earth in the blink of an eye. We were racing in high grass that reminded me of the Everglades as we went by.

"Why are we running so fast?" I asked, but there was no answer. Watching as we moved at high speeds over the land, I was curious as to where we were going. Black Wolf's legs worked fast as I looked down but they were not touching the earth.

Suddenly, I was in the most lush, beautiful rainforest I could have ever pictured and the black wolf was no longer with me. Moving amongst the trees and exotic plants slowly, I could hear the songs from birds, crickets, and frogs and in my mind I asked, "Is this heaven?" Lying before me was a sight that made my heart beat faster. In the small clearing, a coffin was draped as if ready for viewing by a family. My approach was gentle and I hesitated, not sure that I wanted to view the sight but my feet continued to move. Finally, I stood beside the box and looked in. No one was inside and I let out the breath I had been holding. When I looked closer at the reflection in the coffin, I saw the small gold bracelet that I had worn for years laying on the white satin. I picked up the bracelet to look at it closer and without a doubt, it was mine. I laid it back on the satin gently as the world started to spin. I was drifting off into another consciousness. Startled, I opened my eyes to find myself back on my couch.

Picking up pencil and paper, I wrote down everything that I had just experienced and seen on my journey. I saw that an hour had passed when I looked at the clock, but I knew I had not been asleep. Lying back against the pillows, I went over the things that had just happened in my mind. I was confused yet felt refreshed and alive, as if I had been given a great knowledge. A tap on the door scared me and I laughed at myself as I slipped into my gown. Normally, I would have put on my robe or even asked who was at the door, but this time I opened the door without a care.

In the dim light of the hall, I saw a tall figure with an arm full of roses. "Ed," I whispered, feeling relieved as he pulled me into his strong arms. I heard his voice gently saying, "Baby, I love you, you are my world," before he picked me up and carried me to the bedroom, placing me on the waterbed carefully like a sleeping child. He was beside me and we holding each other tight. He took control, moving his hands to touch my face and neck, then using the back of his hand to run down my arm to my hands, entwining his fingers through mine and holding firmly. He kissed my face as if he were trying to devour me. He kissed me deeply until I could no longer breathe.

His seeking mouth caressed my shoulders and his lips traced the outlines of my breasts. His hands holding mine, we moved together like a symphony of beautifully prepared music. I realized I had not said anything during this time. He was in control and I was his to do with as he wanted. I somehow seemed to know when to respond to him, when to kiss him back, as he moved up and down my body with his mouth. Releasing my hands, he lifted my leg and rubbed his face against my calf, moving my body without any trouble, to touch me everywhere. I could feel his breath on my skin as I closed my eyes, lost in the happenings. His hands lovingly explored all my secret places as my body moved with ease to meet his desires. The moments were pure bliss as I coupled my hands around his face, pulling him to my mouth, kissing him. The smell of roses filled the air as we danced the dance of love. Maneuvering as though I were light as a feather, he turned me over, lightly running his fingertips over the lines of my back, kissing me every place his fingers roamed. I put my hand behind me to cover the scars and he kissed my hand, moving it away, placing his cheek in the small of my back on the devastated area. His hands roamed the sides of my frame, pushing up and down the length of my body. He was on his knees between my opened legs, making sure no place went untouched or unloved.

With every kiss, every touch, I felt the ecstatic pleasure of pure love. Pulling my legs up, he turned me again and this time, he faced me as his knees raised my buttocks. I brought myself up, my arms going around him as I buried my face in his chest, listening to the beat of his heart. We fell back against the bed trapped in each other's arms, still kissing as my legs locked his body to mine. Feeling him push as he entered my body, I gasped from the pleasure. My body arched to meet his joining us together as one.

Our movements were like those in a fantasy with no mistakes. I felt him inside my body bringing the hard ravishment to my own being continuously. I took his hand, bringing it up to my lips, kissing his palm, and tasting his fingers. His moans filled the air and I could not escape the enchantment of these actions. Unable to control myself, my hands held his hips as I pushed my body greedily to meet his again and over again. My own senses shook in waves as I felt his body responding. The earth seemed to spin out of control and the stars in heaven soared then exploded on impact with each other. I felt him quivering on me as I tried to breathe again, then he moved to my side, still clutching my body close as we held each other. Drifting off into a peaceful rest, I looked at him and knew there would never be another as my eyes closed.

The next morning, I awoke, startled, and looked to my side—no one was there. The other side of the bed did not appear to have been used, yet I could still feel him with me. What was happening here? Calling out his name, I quickly searched the little apartment, but no one was with me. Pulling on my robe, I ran outside searching for his truck or something and, at this point, I was not sure what I was looking for. The locks on my door were still secure from the inside and that told me no one had left through it since I'd chained it last night. My neighbor Gail was sweeping off her steps as she had done each and every morning for years. Feeling silly, I called to her, asking if she had seen a red truck come in or leave the driveway since last night. Looking around strangely for a second, she answered that she had not seen anyone come in the drive in two days.

Oh my God, I thought. I started shaking as I went back upstairs. Was I losing my mind? Was the night with Ed a delusion or dream? Why did everything seem so real? I kept asking myself. If all that came from smoking a bit of marijuana, then thank God I had not done any LSD or something like that. Feeling lost and confused, I entered my apartment ready to except the dream and the drug possibilities of it all. Looking around my living room, I saw that all the candles had burned out and that the white sage was now ashes in the burner. The scent of the sage still hung heavy in the air. Nothing I saw indicated that he had been with me in any way, shape, or form.

My body felt as though it had been made love to, I thought, as I walked in the bathroom to get a cloth for my face. Looking in the mirror, I could see the redness on my neck and cheeks from whisker burns. Shaking my head as I walked back into my living room, I thought about calling every psychiatrist in the county, maybe even in the state. Holding the cold cloth to my face, I glanced at my coffee table, which had been used for an alter. My heart nearly stopped when I saw a single red rose lying there. My heart starting racing and my whole body trembled as I walked over to look at it closer. I could barely breathe. Slumping down on the edge of the couch, I sat there staring at the rose, wondering if this was also a dream. With my shaking hand, I reached out and touched the delicate petals of a real rose and I knew that this was no dream. The scent of freshness lingered as I picked it up and closed my eyes, inhaling the aroma. Next to the rose was the paper and pencil I had left there with one word on the paper—Ed. Examining it, I knew it was my writing. My heart started pulsating so hard that it shook my body and I went looking for a valium and my nitro tablets. What was happening here? I kept asking myself, but no answers came.

I spent the next couple of days going over the visions of the black wolf, the words from the Old One, and the single red rose. Still wondering about the rose, I had placed it in a vase near my bed. Perhaps I was waiting for it to disappear or to speak, explaining how it came to be.

Knowing I would get a millions calls if I did not explain my absence, I went online long enough to send a few e-mails saying that I was okay, but needed some rest to untangle my thoughts.

I began pacing the floors of my small rooms for hours on end until I was mentally exhausted. I crawled back into my bed to lay there with my eyes wide open thinking for hours at a time. I finally came to the conclusion that I would never be able to explain that rose or how it came into my possession. After examining my thoughts of the dreams or delusions, I was not sure what to call my experience. From what I had discovered on the Internet, I had experienced a "Vision Quest." The purpose of the ritual had been to open up my self-awareness and free my mind and soul. Understanding and comprehension were the key words here, and nobody ever said it would be easy.

I lay on my bed thinking of the black wolf while analyzing each move that had been made in my head and every word that had been spoken in my quest. After thinking for a good while, a few answers became clearer to me. I realized that I lived in a big rush and yet, was going nowhere special. Wanting things now instead of having patience was why he ran so fast with me and never slowed as we traveled. The Old One's words were ones of encouragement, letting me know that I was never alone and that I would always have a greater spirit around me. The coffin with my gold bracelet on the pillow was to tell me that life would go on like the precious metal of gold and even at my death, my memories would never end. The rose was another reminder. Many beautiful things may come into our lives and we must encase each memory in our hearts. When it was gone, we would still have those wonderful images to love and care for when we needed. As for the making love part, I had to assume it was from the drug I had introduced into my system.

On the fifth day after my Quest, I pressed the rose into my journal along with the piece of paper with Ed's name on it and four long, coarse, black hairs I had found on my sheets that were not mine. I had seen enough animal hair to know that if I asked someone for sure, I would be told that they were the hairs from a wolf. There was no need for any further examination on my part, and I closed my journal and placed it away.

All my life, things had been either black or white with a logical explanation behind them. I had never looked deeper into myself and these dreams had been another path for me to open up and search for my truer reality. The possibilities were endless, but one thing was for certain, I felt different inside now. It would be up to me to make the changes to ensure my newfound wisdom would not be wasted but hopefully, it would make the reality of my life stronger and happier. The one feeling that overwhelmed me was that the doctors were wrong, that I would live a long life. Man alone cannot predict death.

I stayed offline for nearly two weeks, which in itself was nothing short of a miracle. I went online only to answer and retrieve e-mail or to talk with Kate and Gwen in voice chat when I saw them. For some reason, I just did not want to deal with all the bullshit any longer. Every time I walked by the computer, I wanted to turn it on and run into my home chat room to wiggle my tail feathers as only the LadyHawk could do. I felt I had come full circle in the past year and before I could go back online, I had to feel secure in myself again. There had been much love, laughter, pain, and disappointment. Maybe the change I wanted so much was due to all the pain I had experienced. Maybe it was the total betrayal I felt I'd received from Ed. I could not be sure, I just knew I had to change my way of thinking if I was going to survive on or offline.

During the next two weeks, I decided to get outside more and went to see many old friends. They all seemed happy to see me but it made me feel bad when I saw the look of pity in their eyes because my legs did not work well anymore. I could only walk short distances before my legs grew weak, and then I would have to sit down to rest. I would tell them there was nothing I could do about that but smile and thank God I could walk at all. It was fun to play pool again, even if I had to hold onto the table before the game was over and I loved to play the jukebox and listen to all the old tunes. Men were always asking me to dance and I wanted to very much, but I knew I could not. My legs would not last a whole song even if it were a slow one.

Some of my friends had bought computers and I began spending a great deal of time setting up their systems. I would go to their homes and teach them basic computer skills and it made me feel good, too. It was fun showing them online games and sites where the kids could play. I was beginning to be happy again and the best things were happening inside me. On other days, Lou and I would go out shopping or to lunch and even to the nearby parks and I could actually feel my body getting stronger at times. More importantly, I felt alive again and my head was becoming clearer and my heart, less heavy.

Jim was calling every other night and we laughed and talked for hours about everything. Mike also called every three days, telling me very interesting stories about the new animals he had seen in the forest where he lived. During all this time, I realized who my real friends were. I discovered I already knew the most important things in my life, but had let them slip away from me somehow. My basic instincts had been correct all along, but I had gotten caught up in a whirlwind of too many other people's lives.

Looking around at my clean house, I knew I had made the right decisions and, since I was on a roll, I decided to call Mike back and invite him down for a home-cooked meal that weekend. When he heard my voice, he laughed and I knew I had been forgiven for canceling our first dinner-date.

I felt more self-confidence now in my abilities to make better choices but regardless, I was still human with human needs. Just as my mind was functioning better, my body was also very alive and yearning. I had examined my online sexual activities as I had everything else in my life, and there were some guilty feelings, but for the life of me, I could not figure out why. There was certainly nothing lewd about masturbation and the men I played with were sweet and kind and I could not see any harm in cybersex. As far as I could see, the only thing that concerned me during the act was my frame of mind. Every one I'd had sex with, online or off, in the last year was always Ed in my mind. I had not made love with anyone without his name running through my head. I had to admit, I had allowed this one-sided love affair to become an obsession in my mind. Maybe that was normal due to my intense feelings for him, but I had gotten too wrapped up in my memories. With my healing phase going on, I had to dig deep within to uncover why I had such a problem letting go.

Chapter 12

The one thing that helped me the most was realizing that my behavior with Ed was normal. I had loved him a great deal and had lost that love and it was much like dealing with a real death of a spouse, a child, or a parent. Death or change is not always easy to deal with where the heart is concerned. I had seen many ladies come into the chat rooms who were dealing with the death of a spouse or a tragic divorce. You could tell they were in a great amount of pain from the words they typed on the screen when talking about their loss. When you feel love inside for another person, it is difficult to say, "Okay, this love is over so I will move on." It does not work like that—at least it doesn't for me. He made it more difficult by coming in and out of the chat rooms. I had made an even bigger mistake in talking to him under the alias of Ed. Yet at that time, I was so stressed over the loss, I would had done anything to keep him close to me even if that meant lying to myself, which I had done so well. At times, I had forgotten that his real name was Larry. It was easy on the computer to reach out and contact him. Would I ever stop loving him? I asked myself. The answer I searched so hard for had been in front of me all along. No, I would never be able to stop my feelings of love for him but I realized that without him returning those feelings, I had nothing. He had committed the ultimate betrayal by sharing my letters with that other woman.

Part of my staying offline had been to put distance between us. Making the adjustments I needed to be a whole person again included my letting go of those memories that had ruled my head and heart. I had to make sure that when I went back online, if I saw him I could stay in control. Even though I knew I was stronger now, I had to know I was going to be okay and face the fact that at some point, his name would light up on my screen again. Feeling renewed in myself and in my faith again, I felt like it was going to be okay. My mind had been made up and set never to allow Larry/Ed to have anymore control over me. I felt free and happy for the first time in many months.

Everything was ready by noon the next day. The food was prepared and the house was clean. Turning the computer on, I went online to check e-mails, then to the chat rooms to look for my friends. Surfing through the rooms, I could see the rooms were much the same but I knew I had changed or maybe I was still in the process of changing. I spoke to a few people in the room, then went back to answering my e-mails. Jim was always "Johnny on the spot" with cards of love and missing me. I thought I had done Jim the same way Larry/Ed had done me in a roundabout way. As I had loved Larry and been rejected, so had Jim loved me and been rejected. The only difference being that I was honest enough to tell Jim my feelings. Perhaps I was too honest with my feelings, but Jim always said that he loved that about me.

Time passed quickly that day and soon I was watching out my window for Mike to arrive. The wind had starting blowing outside and as I stood watching, my mind drifted away to thoughts of spring. I felt lonely listening to the sounds from my wind chimes and the mystical tones almost put me to sleep.

Hearing the roar of Mike's truck pull into the driveway, my eyes flew open to watch his tall, strong frame carrying flowers as he started up my stairs. Waiting until I heard his knock before opening the door, he greeted me with a big smile and a dozen yellow roses. He was so cute. I had never seen him dressed up before and he was very handsome. When we hugged, he smelled delicious, and I grinned but did not let him see. I placed the beautiful flowers into a vase, then thanked him again for his thoughtfulness.

We sat around talking for the longest time while I showed him my family pictures. We were about to eat when I heard car sounds coming up. To my surprise, my boys had decided to pick this time to visit dear old mom. The oldest one came in first and within moments, the youngest came bounding up the stairs. I wondered if they had planned this. Mike stood up to shake hands with both of them and they talked about the weather then about the football games on television. It was strange to have three men in my living room at the same time.

I went into the kitchen to let them talk football, hoping that the boys would soon leave. My oldest followed me into the kitchen smiling, then stood by the counter looking at the food I had prepared. "Mom," he said, "do you have enough for your sons, too?" Pulling him to the side and speaking very low, I said, "I have enough food but this is a date, so you and your brother can eat then get lost." He chuckled and hugged me, going back to the living room. I could hear them talking and laughing as I placed the food on the table making sure it was all set up right.

I did not light the candles before calling to them to come and get it. You would have thought my boys had not eaten in years the way they rushed in. They told Mike I never cooked and that he must be special to get a cooked meal from me. The oldest went to the chair at the head of the table and I put my hand on the back of the chair and gave him a "no" look. He grinned at me and moved to the side chair. Everyone was talking and reaching for the food like it was the Last Supper or something. My youngest decided the candles should be lit without asking me, but he did say, "Hey Mom, you lost your matches for these." I could not help but laugh at him as I glanced at Mike, who was watching me turn many shades of red. My boys were a mess but they were cool at times, too.

151

Dinner went great and lasted longer than I thought it would, with the conversations going on. Finally, my oldest son leans back in his chair rubbing his belly and says, "Well, Mike, what are your intentions with our mother?" I almost went through the floor in embarrassment. That was it, I thought, these boys had provided enough entertainment for one night. Looking at them I said, "If you guys would place your dishes in the kitchen then leave, Mike and I would like to get on with this date." We all laughed out loud. They did take the hint though and said their good nights, continuing to wink and smile at me all the way out the door. They shouted back to me as they left, "Great meal, Mom, glad you have not forgotten how to cook." Oh they were picking at me again but it was fun. I had laughed so much my face was hurting, but I made some fresh coffee before Mike and I settled on the couch. He was a perfect gentleman and I felt at ease with him as we talked about the new wildlife in his area and the baby mammals he was raising.

Everything was almost perfect—great food, good conversation, and a very handsome man. We watched a movie snuggled close on the couch with our feet propped up. Maybe it was all the food and the boring movie, but I found myself waking up sometime later with Mike snoring beside me. He looked comfortable enough so I left him there before climbing into my own bed to fall asleep quickly.

"Wake up," I heard someone say as I was shaken gently. "Hope you don't mind but I took the liberty of fixing breakfast. I am so sorry I fell asleep on you last night," Mike was saying.

Slowly opening my eyes, I could see it was seven a.m. on the clock and I thought, this man must be nuts as I laid back down grumbling to myself about the time. The smell of fresh perked coffee tempted me into rising and going for a shower in record time.

Coming into the living room in my robe as I did each morning, Mike was waiting with a cup of coffee for me plus a big smile. He kept looking at the top of my robe but I just brushed it off, drinking my coffee while it was still hot.

He never said anything about leaving as we talked and I invited him to stay another night if he wanted. The company was good and besides, I had too much food left over from dinner. All that day we watched television and I showed him programs on the computer, teaching him how to set up certain ones. It was a good day and soon we were having a good dinner of leftovers. That man could talk and talk and I was a good listener, but I was getting bored. Mike asked if he could take a shower and I told him where the towels could be found. I heard the water go on and decided to check my e-mail while he bathed. I had no sooner gotten online than I heard Mike call my name asking if I had any shampoo he could use. I told him where the shampoo was, then thought he was like most men since it was right under his nose. I went into the bathroom to point out the secret hiding place of the shampoo in the corner of the tub.

Mike stood by the tub with the towel wrapped around his waist. Holy Mother of God, I thought, what a body. Keeping a straight face, I pointed to the bottle and walked away, closing the door. Without any reservations on my part, I slipped into my little pink gown and turn the CD player on. It was time to see what else was going to happen that night and I was the lady to start it. Hearing the shower stop, I met Mike at the bedroom door holding one of the yellow roses in my hand. He was wearing a tee shirt and sweat pants but looked like a Greek God standing there. Before I could say anything, he smiled and asked if I would like to dance. Moving into his arms, I felt like a schoolgirl, but as soon as his warm hands touched my back, I knew I was all woman. During the whole dance, I don't think we took more than three steps but I knew I was going to be safe in those strong arms, at least for a while.

We stood for seconds, minutes, or perhaps an hour before we unwrapped ourselves from each other. I could not decide in which direction I wanted this relationship to go. On one hand, I wanted to make love to this excellent specimen of man, but on the other hand, I was not sure what I wanted anymore. My knees weakened but Mike noticed, tightening his arms around me. Releasing one arm, he brought his hand up under my chin, gently lifting it up to kiss me. At first, it was a tender, sweet kiss. His warm lips softly pressed on mine, then another followed. With each kiss, his mouth became more demanding. Oh Lord, he could kiss so well, but I was having troubling responding. For a brief instant, I wondered what in the hell was wrong with me. It was unlike me to have any doubts.

Moving my hands from around his back, I slid both arms around his neck, pulling him closer. His mouth was hot against mine and I felt more at ease. He kissed my face, my eyes, and I could feel his body shaking as he moved down to my neck, kissing me passionately. All the time I felt like a performer playing a role, not knowing why I was here, but throwing myself into the part.

We stood there a few seconds longer, then he picked me up and carried me to my bed. He stood tall over me taking only his tee shirt off and lying down beside me. Cuddling in each other's arms, he stroked my hair and continued kissing me. The music played softly in the background as we let our hands caress each other. Without speaking any words, Mike eased my gown off, laying it aside while he greedily cupped and fondled each breast, kissing them as if they were fragile. All the time I was forcing myself to respond to his touches and for the life of me, I did not know why. Closing my eyes, I kept telling myself how good this was and how good Mike was. I rubbed my fingertips in the hair on his chest while his big hands stroked with firm pressure along my back. He whispered to me asking if I would like to have my back rubbed and I said, "Oh yes," turning onto my stomach. The baby oil always stayed on the table beside the bed and I opened it, pouring a little into his hands. The Leo cat was finally coming out in me as I almost purred when he messaged the oil into my back.

Placing the oil back on the table, I could not avoid noticing my journal laying there, too. The fragrance of roses filled the air around me, making me shudder. The scent of the roses was coming through the journal, and I thought about how strange that was. Turning my attention back to Mike, I rolled over and kissed him hard. Damn, I thought, I was not going to let bad memories destroy my whole life. I tried my best to be the woman I knew Mike wanted at this moment.

His hand roamed my back and buttocks gently but firmly, kneading my skin as he whispered how soft I was. He was turning into an aggressive man full of passion driven with desire. He consumed my body with his mouth, kissing me everywhere and I forced myself to play the game. He finally slipped the sweat pants off and we pulled the covers over our heads and giggled like teenagers having a naughty night. Moving my hands to the sides of his face, I guided his mouth down to my breast and gasped as tingles went through my body. Running my hands down his back letting my long nails tease his skin, he continued to ravish my body. His body was wonderful, so strong and solid. There was no part of him that was not giving me everything I should want from a man. He was magnificent in every way. A perfect lover without a doubt, but I could not bring myself to the point of release. The sweat poured from our bodies as Mike did everything possible to help me, but I could not climax. I did something then, that I had never done in my entire life—I faked an orgasm. It was the only way I could put an end to the agony I was putting myself through. Lying in Mike's arms, we whispered tender words and laughed at each other for sweating. I excused myself to the shower and Mike smiled, asking if he could join me. I smiled back telling him that he could be next. Grinning at me, he said okay with a wink.

As the water hit my face, I decided to go back to my psychiatrist. There had to be help for me, a way to live again free. Suddenly, the shower curtain flew open and I screamed. Mike stood there with a shocked look on his face from my startled reaction and he began to apologize. Getting myself under control, I started laughing and grabbed his arm to pull him into the water with me. We held each other and kissed as the water beat against our bodies. Pushing me against the shower wall, he lifted me up and was in me again. Holding my buttocks with both hands, he thrust into me. Within minutes, we were both cumming like crazy. My arms were wrapped around his neck and I kissed his cheeks as he slowly lowered me down. With a look of satisfaction on his face, he kissed me and I was glad he had come into the shower. Maybe I was not as crazy as I had first thought. The water was turning cold and we grabbed our towels, drying off and laughing as we got into our nightclothes.

After we'd dressed, our thoughts turned to a snack. Entering the kitchen, pulling out cold cuts and fruit, we were laughing and acting like we had not eaten in days. I was standing at the counter making a sandwich when Mike looked at me and said, "I noticed the scent of roses in the bedroom. I am so glad you thought to place them in there, it gave the room a nice romantic touch." I almost dropped the lettuce but Mike did not notice. Those roses had not been in that bedroom. In fact, those yellow roses were very beautiful, but had little scent to them. There was no sense in trying to explain all that to Mike and I was not sure I could explain it.

After making my sandwich and helping him assemble his, we went back into the living room to watch a late movie on the television. We drifted off to sleep on the couch, snuggled up in each other's arms.

The next morning right after breakfast, Mike had to go back to Kentucky. He had his animals to feed and other things to do. I think I was ready for him to go, too. He had promised to catch me later on the Yahoo messenger when he was through that day. Before walking to the door, he held me close saying he wanted to come back as soon as he was invited. Smiling up at him I said, "Soon I hope," and he walked out the door.

After pouring another cup of coffee, I went back online. I had missed my computer, which had not been turned on in two days. The first thing I looked for was the self-help online sites. There were many available, and I began to read trying to find out if there was a way I could get over all these anxieties going on inside me. This was a good place to start surfing in my search for an answer.

I had been online only ten minutes when a PM window appeared on my screen with good old Ed wanting to know if I was okay and why I had not been online. Looking at the PM screen, I wondered why it mattered to him. If this were Larry under another username, why did he not say so and stop all the cat and mouse games? What good was coming from all this pretense? Without a word, I closed out the screen as if I had not seen it and continued my surfing. If Ed was Ed and not Larry, I did not care whether I spoke to him again or not. In fact, I was not even sure I wanted to talk to Larry ever again either. Larry's betrayal was so strong in my heart especially with the "she pig" writing the e-mail telling me she'd read all my letters to him. I could not forgive him for that. I think I had been through all the normal emotions over him. Denial, grief, depression, guilt, and now I was working through anger, or trying to. It was difficult with Ed sending PMs and he was a daily reminder of Larry.

Larry or Ms. Piggy had came into the home chat room a few nights back under Larry's username and made the announcement that he was tired of my shit. I was to leave him alone. The trouble was that I had not bothered him in some time. I was just floored by the statement. My friends stood up for me, which made me feel good inside and everyone believe that Ms. Piggy was the one saying all that. Everyone who saw that spectacle had PMed me telling me that it was Ms. Piggy and not Larry—they could tell by the typing technique—but it made no difference to me. We all knew Larry didn't carry the balls in that relationship. I held my cool and acted like a lady, all the time thinking paybacks are hell, then laughing. I knew if he had Ms. Piggy, then the payback was already done. Whatever the pigs had in mind they could leave me out of it. I wasn't lowering myself to their level.

What pissed me off the most was my not understanding why they did this to me or tried to. Public humiliation had not worked. I knew I had come out the winner. Nothing about them mattered to me now. The man I was in love with I felt had died some time ago. I understood that I was in love with the memories of what we'd had. At least that was the attitude I now had about the whole Larry situation.

As I sat there thinking, I received six more PMs and I closed them out and kept surfing. I was not in any mood to play his game since he had upset me again. I was about to close out another PM when I noticed it was from "dewayne2," my new guy. We had spoken several times and even voice chatted but he was still new to me. "Dewayne2" asked to voice chat and I accepted. Looking over the PM window to my chat screen, I noticed the paragraph speaking of healthy sexual feelings. There were no feelings on my part but after what happened with Mike, I wanted to check my sexual deficiencies.

What better way than in cybersex with dewayne2? I thought. I started out very coy in voice playing the part of a naïve woman who had never had cybersex. Men always seemed to be turned on with the thought that they are the first. I flirted, giggled, and made suggestive remarks to "dewayne2" until he was panting all over me in voice chat. After about three minutes, he was putty in my hands. Digging my toy out of its hiding place, I moved to the couch. The sounds of "dewayne2" whispering to me was an instant turn on for me. His soft but firm voice with a strong Texas drawl was just too good. Adjusting my headset, I laid back feeling my nipples getting harder as he spoke of holding me and kissing my neck. The microphone was his for a minute and he was on a roll. My headset was so sensitive, he could actually hear the vibrator humming across my body, and that turned him on even more. Closing my eyes, I thought of his tall frame and the way his hands and mouth would feel on my body.

"I'm kissing your breasts, your nipples getting harder in my mouth as I suck them," he whispered. "Oh yes," I whispered back, "tell me more." His breathing was getting faster as he spoke of caressing my body with one hand while the other continued to fondle each breast. Using the vibrator with much skill, I was beginning to tingle all over. I talked about kissing his neck and nibbling on his ears. Feeling his hot breath against my skin, I made my way down his neck and onto his chest licking, kissing, and lightly sucking his very hard nipples. He groaned as I told him how my soft hands were running up and down his skin playing with his bare back. My fingernails danced over him bringing little goose bumps up all over his naked body. Opening my legs, I let him move on top, still kissing me. His skillful hand was against my hot mound as he played with me. I told him how I was rubbing his ass and moving my hand around, I was now playing with his big, hard dick, stroking it firmly. I told him to rub it against me and asked him if he could feel the steam from my body and he said yes in such a sexy voice that I could have cum right then. We are dancing the dance now, I thought.

He was telling me how he was putting his dick in me very slowly, moving it in and out to open me for him. He was bending over me, kissing me, sucking each breast again and again. Letting the vibrator work around my clit, I was beginning to go nuts listening to him. Our breathing on both ends of the microphone was very fast and he was moaning as I told him I was spreading my legs open further. I wanted him in me now all the way and I was ready. Telling him I was pulling him in closer, I said that I could feel him pumping me even harder. With his naked body hitting mine, thrusting in deeper, his big, hard cock felt so good in me. Both of us must have been ready for this because I could not control myself any longer. His breathing and his gasping voice told me he was ready, too. I whispered to him, "Okay, now, baby, let's do it now." That would be the first and last encounter with "dewayne2."

Lying there, I knew I had nothing but empty feelings. Oh yes, there was a physical satisfaction, but emotionally, there was nothing. Talking into the headset, I told "dewayne2" that he was super, the best ever, and for what it was worth, he was. It was not his fault that I was now so empty inside. Saying our good-byes, we laughed at the wonders of the Internet.

It was a fantasy, I thought, lying there thinking about things. All of a sudden, a cloud seemed to be lifted from me and I knew what I wanted to do. It was like a revelation of the new things I could do or wanted to do with my life, like bits and pieces of the puzzle coming together. Taking a deep breath, I tried to keep my thoughts organized. Grabbing a pencil and paper, I began figuring money and taking stock of all my possessions. For the first time in years, it was clear to me what would make me happy and give me peace in the world.

Chapter 13

After writing and thinking of the possibilities, I decided it would be best not to rush into anything for if my plans for a new life were going to work, I had to make certain that I made no mistakes. I had to make certain I was going to be happy with my decisions also. On the top of the paper I wrote, "What I really want." A simple title but to the point. Logging back onto the computer, I ran several searches on rental properties in the mountains of east Tennessee. To find what I was looking for would require a visit up to that area and I still had so much to think over. The feeling of total peace washed through me. That peaceful feeling was like a sign that I was now on the right path for me and I prayed for peace and guidance about the matter. For the next week, I wrote down everything I felt and took inventory on everything I owned, which was not much.

I did have a few things that were considered antiques. They were not worth a great deal, but I needed to sell everything to make this change. Looking around me, all I saw were bad memories and useless items collecting dust. I felt certain that if I wanted major changes I would have to be the one to make them happen. Picking up the phone, I called Lou asking her if she would like to come over for some coffee—that I needed to talk with her. She was busy at the moment, but promised to come over later. Next I called Ken up in Kentucky and asked him about me helping with raising the baby animals again in the spring. I had done this before a few years back when I lived near him, so I did have the knowledge of how to handle the babies that would possibly come to me. He was delighted with the ideal and said there was another wildlife center I could contact that might also need my help raising the baby animals. Ken had been my wildlife teacher and had taught me all I needed to know about nursing the very young mammals. He was always on standby with answers to all the questions I could ever have. If he did not know, he would know where to find the answer. Before my computer had taken over my life, I had actually studied about wildlife. When I turned on the computer at first, I'd spent countless hours at Discovery.com learning all I could. Suddenly, life seemed to have meaning again. All my thoughts of the future were feasible and all I had to do was keep working toward the goals I wanted. Making a checkmark by Ken's name, I smiled because with his backing as a caretaker, I could have people calling me wanting my help. There was never a charge for these types of volunteer work, just personal satisfaction, which was fine with me. My excitement grew. That part was settled.

For the next three weeks, I went through my tons of boxes I had stored. The boxes contained the things I had gathered up over the years—tons of dishes, knick-knacks, and clothes that I could not bear to throw away. I thought maybe one day I would want or need them. Now I was organizing the boxes and placing strips of white tape on everything with garage sale prices. At night, I was online surfing the chat rooms. In my home chat room, I sat back watching the names come on the screen as I had done so many times. Ed was still sending PMs and I talked to him but my feelings had changed so much. It no longer mattered to me if Ed was Larry and whatever I had felt for Larry was now over. My dear friend Kate said that was good that I felt like that, because I could now get on with my life. I knew she was right, too. Everyday I saw the idiot who had sent the e-mail telling me she and Larry were now together. Even in my mellow state of mind, I would always hate that bitch for sending that e-mail. Still, I believed that karma would slap them both down. What goes around, comes around. I would let that old saying stick in my mind where they were concerned from now until forever.

Jim continued calling every other night and I told him of my plans and he was upset, but said he understood. Mike and I still talked on the Yahoo messenger every day and he was always wanting to come back to visit me, but my heart was now closed, and I was not sure when it had sealed itself. My thoughts of finding love were now closed after the pain and hurt from Larry. I could no longer face the possibilities of ever loving again and that made my decisions easier to work out. During the next two weeks, I had garage sales that made a fairly decent amount of money and I had gotten rid of all the extra junk I had collected.

My children were the last to know of my plans but one Sunday, I called the boys over and told them what I was about to do. At first, they were upset but they knew me well enough to know that I was going to do what I wanted. They were old enough to live their lives without Mom to handle their daily troubles. They were men and had their own lives as I had mine. My daughter was the one who had helped me the most of my kids with her moral support. Talking to her on the phone one day she said, "Mom, I love you and if you think you will be happy then you go do whatever you need to do." No one could have a better daughter in the world, I was so proud of her, and felt lucky that she was mine.

I had placed ads in Yahoo classifieds stating what I was looking for and where and had received a few great answers, too. Everything was working out exactly the way I wanted. Things were going so smooth I knew I was doing the right thing. I had one more garage sale then all I had left were a few dishes, my television, and computer. Looking around my small cave that had been my home for so long, I knew I would miss it but the excitement for my future was overpowering. I always knew there was that tiny flicker in me and I could change my mind at any time.

It must have been a tiny flicker because I never hesitated once I started.

Heading back online, I sent e-mails to everyone I considered a friend asking for an address to write them at. Some responded, some did not, and that was okay, too. I called Ken and told him I would see him in a couple of days and asked if he could keep Rebel for a week while I made all the final arrangements and set up plans. I knew he would help with anything I asked of him within his power. Kate and Gwen knew everything and stood behind me like true friends do. Lou was with me all the way, besides, she liked to travel and coming to visit me would give her something to look forward to.

I was just about to go offline when I received a PM from Ed asking me where I was going and what I was doing. He had been one I asked for an address to write to but did not receive an answer. I simply stated that everything changes in life, that nothing remains the same and after clicking the send button, I closed out the PM screen. Saying my silent farewells to him, I closed down my computer, unplugged it, and carried it out to the car. I picked up my journal and the scent of that one red rose was still so overpowering, I did not think I could ever write in it again. In fact, I had not written anything in it since the night the rose appeared to me.

Early the next morning, I took one last look at my apartment and remembered some of the good things I had learned while living there. I loaded Rebel into the car and threw my journal on the back seat. Pulling out of the drive, I felt a little sad but I knew my future was just up the road. I had a long way to go and I understood it would not be easy, but nothing worth having in life is ever easy.

Arriving at Ken's a few hours later, I watched Rebel playfully run through the woods and fields. Sitting there having one cup of coffee with Ken, I knew I had things I had to do and a long way to go. Hugging Rebel and telling him I would see him soon, I jumped back into the car and headed out. I did have one stop to make before leaving Kentucky. Going to that beautiful, lonely place by the Tennessee River, I pulled my car to the edge of the water.

I sat on the riverbank for an hour watching the water run swiftly by, feeling the strength of the river. I opened the door to my car and, picking up my computer, walked to the edge, tossing it as hard as I could out into the water. It was my way of saying good-bye to the world I had known. I took my journal and walked to the grassy part of the grove that Ken said he thought ancient people had once used as a gathering place and buried my journal in a small, shallow grave. I cried so hard while standing there, I grew weak and fell to my knees. I prayed over the things I had lost and held dear to my heart and prayed for the strength to follow my new path. Finally, there were no more tears, and pulling myself up, I looked around and said my final good-byes. The feelings were intense within me and I felt as though I had just been to a funeral. In many ways, I had. Getting back into the car, I drove to my future. I would never look back in regret ever again.

(Three months later)

Grabbing my coffee cup, I open the front door and Rebel greets me as he bounces around in front of me. The sun is coming up over the mountains and the mist in the valley below is breathtaking. Sitting in my porch swing, I smell the freshness of the new day while I sip my coffee. Rebel lies at my feet listening and watching every movement around us. Looking over my shoulder at my new garden, I see the plants breaking through the ground and all is well with my world.

Checking my real mailbox, there was a letter from Ken. In the letter he said he had been to the river where I had thrown my computer and much to his surprise, there was now a beautiful rose bush growing there. Right in the middle of that grassy spot where I had buried my journal. Smiling as I read further. He asked if I had planted any seeds when I was out there with a big question mark. I would answer him on that someday, I thought. Feeling contentment in all my labors, I heard the whining of new fox pups wanting their morning meal, too. Taking in a deep breath, I thanked God for everything I had and for my new life. Spring has arrived and everything is waking up and living again—and so am I.